Water waves and ship hydrodynamics

MECHANICS OF FLUIDS AND TRANSPORT PROCESSES
Editors: R.J. Moreau and G.Æ. Oravas

J. Happel and H. Brenner, Low Reynolds number hydrodynamics. 1983. ISBN 90-247-2877-0.
S. Zahorski, Mechanics of viscoleastic fluids. 1982. ISBN 90-247-2687-5.
J.A. Sparenberg, Elements of hydrodynamic propulsion. 1984. ISBN 90-247-2871-1.
B.K. Shivamoggi, Theoretical fluid dynamics. 1984. ISBN 90-247-2999-8.
R. Timman, A.J. Hermans and G.C. Hsiao, Water waves and ship hydrodynamics: An introduction. 1985. ISBN 90-247-3218-2.

Water waves and ship hydrodynamics

An introduction

By

R. Timman†

A.J. Hermans

Department of Mathematics
Delft University of Technology
Delft, The Netherlands

G.C. Hsiao

Department of Mathematics
University of Delaware
Newark, DE 19711, USA

SPRINGER-SCIENCE+BUSINESS MEDIA, B.V. 1985

Library of Congress Cataloging in Publication Data

Timman, R. (Reinier), 1917-1975.
Water waves and ship hydrodynamics.

(Mechanics of fluids and transport processes ; 5)
Bibliography: p.
Includes index.
1. Water waves. 2. Ships--Hydrodynamics.
I. Hermans, A. J. II. Hsiao, G. C. (George C.)
III. Title. IV. Series: Mechanics of fluids and
transport processes ; v. 5)
TC172.T56 1985 623.8'144 85-18807

ISBN 978-90-481-8292-3 ISBN 978-94-017-3657-2 (eBook)
DOI 10.1007/978-94-017-3657-2

CONTENTS

PREFACE

In the spring of 1971, Reinier Timman visited the University of Delaware during which time he gave a series of lectures on water waves from which these notes grew. Those of us privileged to be present during that time will never forget the experience. Rein Timman is not easily forgotten.

His seemingly inexhaustible energy completely overwhelmed us. Who could forget the numbing effect of a succession of long wine filled evenings of lively conversation on literature, politics, education, you name it, followed early the next day by the appearance of the apparently totally refreshed red haired giant eager to discuss our mathematical problems with keen insight en remarkable understanding, ready to lecture on fluid mechanics or optimal control theory or a host of other subjects and ready to work into the evening until the cycle repeated. He thought faster, he knew more, he drank more and he slept less than any of us mortals and he literally wore us out. What a rare privilege indeed to have participated in this intellectual orgy. Timman's lively interest in almost everything coupled with his buoyant enthusiasm and infectious optimism epitomized his approach to life. No delicate nibbling at the fringes, he wanted every morsel of every course.

In these times of narrow specialization truly renaissance figures are, if not extinct, at least a highly endangered species. But Timman was one of that rare breed. His knowledge in virtually all areas of classical applied mathematics was prodigious. I still marvel that while I was his doctoral student in Delft in the late fifties working on a problem in electromagnetic scattering he also had students working in water waves, cavitation, elasticity, aerodynamics and numerical analysis. He was a boundless source of inspiration to his students in all of these varied fields.

His inattention to detail is legendary but this did not hamper his ability to focus on what was really important in a problem. With a wave of his large hand he would dismiss unimportant errors while concentrating on central ideas, leaving to us the task of setting things right mathematically. This nonchalant attitude toward minus signs and numerical factors was probably deliberate. He wanted people to see the forests, not the trees; to focus on the heart of the problem not inconsequential superficialties. He had little use for the all too prevalent penchant for examining someone's work looking for errors. He would read a paper looking for the gold, not the dross; looking for what was right, not what was wrong.

Of course this did not make life easy for those around him but it did make it interesting. This will be attested to by George Hsiao and Richard Weinacht whose revised version of the notes from Timman's water wave lectures appeared as a University of Delaware report. Timman and Hsiao then planned to further revise and expand these notes and publish them in book form but the project came to an abrupt halt with Reinier Timman's untimely death in 1975. It might have remained unfinished had not Aad Hermans' visit to Delaware in 1980 breathed new life into it. Together George Hsiao and Aad Hermans have completed the task of revising the notes, reorganizing the presentation, restoring the factors of two which Timman had cavalierly omitted, and adding some new material. The first four chapters are based substantially on the original notes while the fifth chapter and the appendices have been added.

It is gratifying to see the completion of these notes. Hopefully they will provide a useful introduction to water waves for a new generation of mathematicians and engineers. This area was perhaps first among equals in the broad spectrum of Timman's interests. If these notes succeed in stimulating this new generation to concentrate on the challenging problems remaining in this field they will serve as a fitting memorial to a remarkable man whose like will not be soon seen again.

March, 1985

Ralph E. Kleinman
Newark, Delaware

ACKNOWLEDGEMENTS

The authors wish to thank Professor Willard E. Baxter for his active interest in the publication of these lecture notes. We are grateful to Professor Richard J. Weinacht who spent a great deal of time helping one of us (GCH) in preparing the original draft of the water wave notes on which the present first four chapters were based. We owe a special debt to Professor Ralph E. Kleinman without whom this project probably would never have started.

We would also like to express our appreciation to Mr. G. Broere for making most of the drawings, and to Mrs. Angelina de Wit for typing the final manuscript.

Finally, we would like to express our gratitude to the Department of Mathematical Sciences of the University of Delaware, the Department of Mathematics and Informatics of the Delft University of Technology and the Alexander von Humboldt Stiftung for financial support at various stages of the preparation of this manuscript.

March, 1985 AJH and GCH

CHAPTER I - LINEAR THEORY OF WATER WAVES

This chapter contains the formulation of boundary and initial-boundary value problems in water waves. The basic equations here are the Euler equations and the equation of continuity for a non-viscous incompressible fluid moving under gravity. Throughout the book, in most considerations the motion is assumed to be irrotational and hence the existence of a velocity potential function is ensured in simply connected regions. In this case the equation of continuity for the velocity of fluid is then reduced to the familiar Laplace equation for the velocity potential function.

Water waves are created normally by the presence of a free surface along which the pressure is constant. For the irrotational motion, on the free surface one then obtains the non-linear Bernoulli equation for the velocity potential function from the Euler equations. Based on small amplitude waves, linearized problems for the velocity potential function and for the free surface elevation are formulated [8], [12].

1.1 Basic Equations

The theory of water waves, to be presented here, is based on a model of non-viscous incompressible fluid moving under gravity. The equations of motion will be expressed in a right handed system of rectangular coordinates x, y, z. In the Euler representation they read:

$$\begin{aligned} u_t + uu_x + vu_y + wu_z &= -\frac{1}{\rho} P_x, \\ v_t + uv_x + vv_y + wv_z &= -\frac{1}{\rho} P_y - g, \\ w_t + uw_x + vw_y + ww_z &= -\frac{1}{\rho} P_z. \end{aligned} \tag{1.1}$$

Here $u = u(x,y,z,t)$, $v = v(x,y,z,t)$, $w = w(x,y,z,t)$ are velocity components in the corresponding x,y,z direction; $p = p(x,y,z,t)$ is the pressure; ρ is the density of fluid, a constant, and g is the gravitational acceleration. The continuity equation is:

$$u_x + v_y + w_z = 0. \tag{1.2}$$

In most of the considerations the motion is considered to be *irrotational*. This gives the additional set of equations

$$\begin{aligned} u_y - v_x &= 0, \\ v_z - w_y &= 0, \\ w_x - u_z &= 0, \end{aligned} \tag{1.3}$$

which guarantees in a simply connected region the existence of a velocity potential φ with

$$\begin{aligned} u &= \varphi_x, \\ v &= \varphi_y, \\ w &= \varphi_z. \end{aligned} \tag{1.4}$$

From equation (1.2) we see that φ satisfies Laplace's equation

$$\varphi_{xx} + \varphi_{yy} + \varphi_{zz} = 0. \tag{1.5}$$

This greatly facilitates the theory.

In general, however, solutions of Laplace's equation will not show wave character, since the equation is elliptic. Waves are created by the presence of a free surface and are intimately related to the surface condition.

1.2 Boundary Conditions

At a moving boundary the condition for a non-viscous fluid is very simple. The fluid velocity normal to the surface has to be equal to the normal component of the velocity of the surface itself. If the equation of the surface is given by

$$y = F(x,z,t), \tag{1.6}$$

we denote the velocity of a point on the surface by (U, V, W). A normal to the surface has the direction cosines

$$\left(\frac{F_x}{\sqrt{F_x^2 + F_y^2 + 1}}, \frac{-1}{\sqrt{F_x^2 + F_y^2 + 1}}, \frac{F_z}{\sqrt{F_x^2 + F_y^2 + 1}}\right) \tag{1.7}$$

and the surface (or boundary) condition reads

$$\frac{uF_x + wF_z - v}{\sqrt{F_x^2 + F_y^2 + 1}} = \frac{UF_x + WF_z - V}{\sqrt{F_x^2 + F_y^2 + 1}} = \frac{-F_t}{\sqrt{F_x^2 + F_y^2 + 1}}, \tag{1.8}$$

because

$$F_xU - V + F_zW + F_t = 0$$

for a point on the moving surface. Hence from (1.8) we have

$$v = F_t + uF_x + wF_z = \frac{dF(x,z,t)}{dt}, \tag{1.9}$$

which expresses the fact, that once a fluid particle is on the surface, it remains on the surface.

We will usually denote the bottom surface by $y = H(x,z,t)$, so that equation (1.9) reads

$$v = H_t + uH_x + wH_z. \tag{1.10}$$

Mostly in our considerations the bottom is fixed, that is H is independent of t, so that the term H_t in (1.10) vanishes.

The waves are created at the free surface, which is characterized by the condition that along this surface the pressure is a constant. Hence in addition to the kinematic equation

$$v = \eta_t + u\eta_x + w\eta_z, \tag{1.11}$$

for the free surface $y = \eta(x,z,t)$, we have the equation

$$p = \text{constant}, \tag{1.12}$$

along $y = \eta(x,z,t)$. There are two ways of formulating these conditions:

a. From the equations of motion (1.1), we find by inspection, in the case of irrotational motion, the Bernoulli equation

$$\varphi_t + \frac{1}{2}\left(u^2 + v^2 + w^2\right) + gy + \frac{p}{\rho} = f(t) \tag{1.13}$$

which, because of the constant pressure, one can always normalize φ to result in the dynamical free surface condition

$$\varphi_t + \frac{1}{2}\left(\varphi_x^2 + \varphi_y^2 + \varphi_z^2\right) + g\eta = \text{constant}. \tag{1.14}$$

b. The second way expresses that

$$\frac{\partial p}{\partial s_x} = 0, \qquad \frac{\partial p}{\partial s_z} = 0, \tag{1.15}$$

where s_x and s_z are coordinates on the free surface, which have their projections in the x and z directions, respectively. This gives*)

$$\begin{aligned} \frac{\partial p}{\partial s_x} &= \frac{\partial p}{\partial x} \cos(x,s_x) + \frac{\partial p}{\partial y} \cos(y,s_x) = 0, \\ \frac{\partial p}{\partial s_z} &= \frac{\partial p}{\partial z} \cos(z,s_z) + \frac{\partial p}{\partial y} \cos(y,s_z) = 0 \end{aligned} \tag{1.16}$$

*) Note that $\cos(x,s_z) = \cos(z,s_x) = 0$.

or

$$p_x + p_y\eta_x = 0,$$
$$p_z + p_y\eta_z = 0. \qquad (1.17)$$

Substituting equation (1.17) into (1.1), we have the relations

$$u_t + uu_x + vu_y + wu_z + \eta_x(v_t+uv_x+vv_y+wv_z+g) = 0,$$
$$w_t + uw_x + vw_y + ww_z + \eta_z(v_t+uv_x+vv_y+wv_z+g) = 0, \qquad (1.18)$$

which are also valid for rotational flow.

In this way the basic equations are derived. The further development of the theory is based on small parameter expansions of these equations.

1.3 Linearized Theory (small amplitude waves)

The simplest approximation is the case, where the deviation η of the free surface above a certain standard level which is taken as $y = 0$ is small. We assume that

$$\eta(x,z,t) = \varepsilon\bar{\eta}(x,z,t), \qquad (1.19)$$

where ε is a small parameter. In addition, we assume the bottom slope to be small of the same order of magnitude in ε, and put

$$y = -h + \varepsilon h_1, \qquad (1.20)$$

which will lead to the boundary condition

$$v = \varepsilon(h_{1t} + uh_{1x} + wh_{1z}) \qquad (1.21)$$

from equation (1.10). For the free surface we obtain from equations (1.11) and (1.14)

$$v = \varepsilon(\bar{\eta}_t + u\bar{\eta}_x + w\bar{\eta}_z) \qquad (1.22)$$

together with

$$\varphi_t + \frac{1}{2} (\varphi_x^2 + \varphi_y^2 + \varphi_z^2) + \varepsilon g\bar{\eta} = \text{constant}. \tag{1.23}$$

Now, for the solution of equation (1.5), we assume an expansion

$$\varphi(x,y,z,t) = \varphi_0 + \varepsilon\varphi_1 + \varepsilon^2\varphi_2 + \ldots\ldots, \tag{1.24}$$

and substitute in the equation (1.5) and boundary condition (1.21). Equating to zero the coefficients of like powers of ε, we get first that all φ_k's are harmonic functions. Moreover, we have from equations (1.20) and (1.21)

$$\left.\begin{aligned} v_0 &= \varphi_{0y} = 0 \\ v_1 &= h_{1t} + u_0 h_{1x} + w_0 h_{1z} \end{aligned}\right\} \text{ at } y = -h. \tag{1.25}$$

Similarly, we expand $\bar{\eta}$ in (1.19) in the form

$$\bar{\eta} = \eta_1 + \varepsilon\eta_2 + \varepsilon^2\eta_3 + \ldots\ldots, \tag{1.26}$$

and we find from equation (1.22) the free surface condition at $y = \varepsilon\bar{\eta}$

$$\begin{aligned} v_0 &= 0 \quad \text{and} \\ v_1 &= \eta_{1t} + u_0\eta_{1x} + w_0\eta_{1z} \end{aligned} \tag{1.27}$$

together with

$$\begin{aligned} \varphi_{0t} + \frac{1}{2}(\varphi_{0x}^2 + \varphi_{0y}^2 + \varphi_{0z}^2) &= \text{constant}, \\ \varphi_{1t} + u_0u_1 + v_0v_1 + w_0w_1 + g\eta_1 &= 0 \end{aligned} \tag{1.28}$$

from equation (1.23).

The first approximation φ_0, $u_0 = \varphi_{0x}$, $v_0 = \varphi_{0y}$, $w_0 = \varphi_{0z}$, corresponds to a permanent flow. If we take the special case:

$$u_0 = \text{constant},$$
$$v_0 = 0,$$
$$w_0 = \text{constant},$$

we can transform to a coordinate system with the x-axis in the direction of this constant flow and denote the velocity by U. In this case we have $\varphi_0 = Ux$ and the constant in (1.23) is equal to $\frac{1}{2}U^2$. Then we have the boundary condition from (1.25)

$$v_1 = Uh_{1x} \quad \text{at } y = -h \tag{1.29}$$

and at the free surface $y = \varepsilon\bar{\eta}$, the coefficient of ε (or equations (1.27) and (1.28)) gives

$$\begin{aligned} \varphi_{1y} &= \eta_{1t} + U\eta_{1x}, \\ \varphi_{1t} + U\varphi_{1x} + g\eta_1 &= 0. \end{aligned} \tag{1.30}$$

Instead of putting this condition (1.30) at $y = \varepsilon\bar{\eta}$, we put it at $y = 0$. Assuming that φ_1 admits an expansion in powers of $\bar{y}$, we have then

$$\begin{aligned} \varphi_{1x}(x,\varepsilon\bar{\eta},z) &= \varphi_{1x}(x,o,z) + \varepsilon\bar{\eta}\varphi_{1xy}(x,o,z) + \ldots\ldots \\ &= \varphi_{1x}(x,o,z) + \varepsilon\eta_1\varphi_{1xy}(x,o,z) + 0(\varepsilon^2), \end{aligned}$$

which leads to a modification of the terms of second order or higher. Hence the first approximation gives the following set of linear equations for φ_1 and $\bar{\eta}_1$:

$$\begin{aligned} &\varphi_{1xx} + \varphi_{1yy} + \varphi_{1zz} = 0 \\ &\varphi_{1y} = h_{1t} + Uh_{1x} \qquad \text{at } y = -h, \\ &\left.\begin{aligned} \varphi_{1y} &= \eta_{1t} + U\eta_{1x} \\ \varphi_{1t} + U\varphi_{1x} + g\eta_1 &= 0 \end{aligned}\right\} \text{ at } y = 0. \end{aligned} \tag{1.31}$$

For a fixed flat bottom, h_1 is constant so that $h_{1x} = h_{1t} = 0$. For smooth functions, one can easily eliminate η_1 in the surface condition and obtain the formulation for the first order approximation (dropping subscript 1):

$$\begin{aligned} &\varphi_{xx} + \varphi_{yy} + \varphi_{zz} = 0 \\ &\varphi_y = 0 && \text{at } y = -h \\ &U^2\varphi_{xx} + 2U\varphi_{xt} + \varphi_{tt} + g\varphi_y = 0 && \text{at } y = 0. \end{aligned} \tag{1.32}$$

Here the surface elevation η can be computed by

$$\eta = \frac{-1}{g}(\varphi_t + U\varphi_x). \tag{1.33}$$

Now given initial conditions, problems as defined by (1.32) can be solved by means of Laplace or Fourier transform method. As for illustration, we shall consider a few simple examples in Chapter 2.

CHAPTER II - LINEAR WAVE PHENOMENON

A few simple examples of the linearized boundary and initial-boundary value problems formulated in the previous chapter will be solved by the Fourier or Laplace transform method. Through these simple examples, basic wave phenomena or terminologies in water waves will be introduced. These are *phase velocity* (§2.2), *dispersion relation* (§2.3), *group velocity* (§2.4), *wave fronts* (§2.5), to name a few.

Of particular importance is the asymptotic behaviour of the free surface elevation for large values of relevant spaces and for time variables. This behaviour can be best obtained by the method of stationary phase (see (A.1)). In this connection, the method of characteristics for treating first order non-linear partial differential equations for the phase function is employed. Hence a brief summary of the concepts of characteristics is also included in the Appendix (A.2).

2.1 The Moving Pressure Point

For small amplitude waves the linearized problem is defined by (1.32). We suppose the bottom to be at infinity. In this case one has to replace the boundary conditions in (1.32) by the condition that φ is finite as $y \to -\infty$. We look for a very simple solution in a steady flow, for which everywhere at $y = 0$ except at $x = z = 0$ the pressure vanishes. By introducing the dimensionless coordinates:

$$\bar{x} = \frac{xg}{U^2}, \quad \bar{y} = \frac{yg}{U^2}, \quad \bar{z} = \frac{zg}{U^2},$$

we can formulate the boundary value problem as follows:

$$\varphi_{\bar{x}\bar{x}} + \varphi_{\bar{y}\bar{y}} + \varphi_{\bar{z}\bar{z}} = 0$$

$$\varphi_{\bar{x}\bar{x}} + \varphi_{\bar{y}} = 0 \quad \text{at } \bar{y} = 0,\ \bar{x} \neq 0,\ \bar{z} \neq 0 \tag{2.1}$$

$$\varphi \text{ finite} \quad \text{as } y \to -\infty.$$

We seek a solution of (2.1) in the form of Fourier integral in $\bar{x}$,

$$\varphi(\bar{x},\bar{y},\bar{z}) = \frac{1}{2\pi} \int_{-\infty}^{\infty} e^{-i\alpha\bar{x}}\, \tilde{\varphi}(\alpha,\bar{y},\bar{z})\, d\alpha, \tag{2.2}$$

where $\tilde{\varphi}(\alpha,\bar{y},\bar{z}) = \int_{-\infty}^{\infty} e^{i\alpha\bar{x}}\, \varphi(\bar{x},\bar{y},\bar{z})\, d\bar{x}$. This leads to the boundary value problem for $\tilde{\varphi}(\alpha,\bar{y},\bar{z})$:

$$\tilde{\varphi}_{\bar{y}\bar{y}} + \tilde{\varphi}_{\bar{z}\bar{z}} - \alpha^2\tilde{\varphi} = 0$$

$$-\alpha^2\tilde{\varphi} + \tilde{\varphi}_{\bar{y}} = 0 \quad \text{at } \bar{y} = 0 \tag{2.3}$$

$$\tilde{\varphi} \text{ finite} \quad \text{as } \bar{y} \to -\infty.$$

A simple solution of (2.3) can be found in the form

$$\tilde{\varphi} = e^{\alpha^2\bar{y}}\, F(\bar{z}),$$

where $F(\bar{z})$ satisfies the equation

$$(\alpha^4 - \alpha^2)F + F_{\bar{z}\bar{z}} = 0.$$

Consequently, we take as a possible solution

$$\varphi(\bar{x},\bar{y},\bar{z}) = \frac{A}{2\pi} \int_{-\infty}^{\infty} \exp\left\{-i\alpha\bar{x} + \alpha^2\bar{y} + i\alpha(\alpha^2 - 1)^{\frac{1}{2}}\, \bar{z}\right\} d\alpha, \tag{2.4}$$

for A being a constant. Note that $\varphi(\bar{x},\bar{y},\bar{z})$ in (2.4) is not defined for $\bar{x} = \bar{y} = \bar{z} = 0$. From (1.33) we find the free surface elevation

$$\eta(\bar{x},\bar{z}) = \frac{Ai}{2\pi U} \lim_{\bar{y}\to 0} \int_{-\infty}^{\infty} \left(\alpha e^{\alpha^2 \bar{y}}\right) \exp\left\{i\left(-\alpha\bar{x} + \alpha(\alpha^2 - 1)^{\frac{1}{2}} \bar{z}\right)\right\} d\alpha \tag{2.5}$$

which apparently is infinite for $\bar{x} = \bar{z} = 0$.

In order to get a better insight into the shape of the surface we shall evaluate this expression (2.5) for large values of $\bar{x}$ and $\bar{z}$; that is, distances to the origin which are large compared to the reference length U^2/g. This evaluation is performed by the method of stationary phase (see (A.1)).

We note that if we let $r = (\bar{x}^2 + \bar{z}^2)^{\frac{1}{2}}$, $\bar{x} = r \cos \vartheta$ and $\bar{z} = r \sin \vartheta$, then for each fixed ϑ, (2.5) can be written in the form

$$\int_{-\infty}^{\infty} g(\alpha) \exp\left\{irf(\alpha)\right\} d\alpha,$$

where

$$g(\alpha) := \frac{Ai}{2\pi U} \alpha \quad \text{and}$$

$$rf(\alpha) := -\alpha\bar{x} + \alpha(\alpha^2 - 1)^{\frac{1}{2}} \bar{z}.$$

Hence the stationary points are solutions of the equation:

$$\frac{\partial}{\partial\alpha}\left\{-\alpha\bar{x} + \alpha(\alpha^2 - 1)^{\frac{1}{2}} \bar{z}\right\} = 0. \tag{2.6}$$

(cf. (A.1)).

Let α_0 be a solution of (2.6). We obtain therefore the asymptotic form of $\eta(\bar{x},\bar{z})$:

$$\eta(\bar{x},\bar{z}) \cong \frac{Ai}{\pi U} \alpha_0 \sqrt{\frac{\pi i \alpha_0 (\alpha_0^2 - 1)^{3/2}}{2\bar{z}(2\alpha_0^2 - 3)}} \exp\left\{-\alpha_0\bar{x} + \alpha_0(\alpha_0^2 - 1)^{\frac{1}{2}} \bar{z}\right\}. \tag{2.7}$$

The phase of the wave is of the most importance. If we put

$$\psi = -\alpha_0\bar{x} + \alpha_0 \sqrt{\alpha_0^2 - 1}\, \bar{z}, \tag{2.8}$$

we obtain from (2.6)

$$-\bar{x} + \frac{2\alpha_0^2 - 1}{\sqrt{\alpha_0^2 - 1}}\,\bar{z} = 0. \qquad (2.9)$$

Setting $\alpha_0 = \frac{1}{\cos\vartheta}$, we obtain from (2.8) and (2.9) the equations:

$$\begin{aligned} \bar{x} &= -\psi\,(2\cos\vartheta - \cos^3\vartheta) = -\tfrac{1}{4}\psi\,(5\cos\vartheta - \cos 3\vartheta) \\ \bar{z} &= -\psi\,(\cos^2\vartheta\,\sin\vartheta) = -\tfrac{1}{4}\psi\,(\sin\vartheta + \sin 3\vartheta) \end{aligned} \qquad (2.10)$$

for the curves of constant phase ψ, which give the wave pattern. These curves are all similar with the origin as center, and have cusps at $\bar{x} = \bar{z} = 0$ (or $\vartheta = \pi/2$) and at the points where $d\bar{x}/d\vartheta = d\bar{z}/d\vartheta = 0$. Since $d\bar{x}/d\vartheta = -\psi\sin\vartheta\,(2 - 3\cos^2\vartheta)$ and $d\bar{z}/d\vartheta = -\psi\cos\vartheta\,(3\cos^2\vartheta - 2)$, it follows that at the points

$$\bar{x} = -\psi\,\frac{4\sqrt{6}}{9}, \qquad \bar{z} = -\psi\,\frac{2\sqrt{3}}{9}$$

which correspond to $\cos\vartheta = \sqrt{2/3}$, there are cusps (it is understood that the expression (2.7) is not valid in the neighbourhood of cusps). A typical curve is shown in Figure 2.1. We see that the curve intersects the $\bar{x}$-axis at the point $\bar{x} = -\psi$ (corresponding to $\vartheta = 0$). These cusps lie on a straight line, through the origin, which makes a fixed angle with the $\bar{x}$-axis. The pattern obtained this way is called the Kelvin Wave pattern.

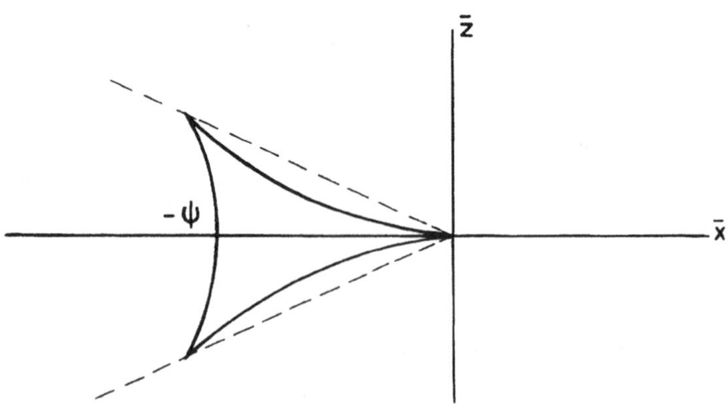

Figure 2.1

2.2 Oscillating Singularity

The boundary value problem for an oscillating singularity at rest at the origin is defined by the same equations in (1.32) for small amplitude waves except that in this case U = 0. For harmonic oscillations we put

$$\begin{aligned} \varphi(x,y,z,t) &= \hat{\varphi}(x,y,z)\, e^{i\omega t}; \\ \eta(x,z,t) &= \hat{\eta}(x,z)\, e^{i\omega t}. \end{aligned} \tag{2.11}$$

Then the amplitude function $\hat{\varphi}(x,y,z)$ satisfies the Laplace equation

$$\hat{\varphi}_{xx} + \hat{\varphi}_{yy} + \hat{\varphi}_{zz} = 0 \tag{2.12}$$

and the surface condition

$$-\omega^2\hat{\varphi} + g\hat{\varphi}_y = 0 \qquad \text{for } y = 0. \tag{2.13}$$

For infinite depth, we have again the condition

$$\hat{\varphi} \text{ finite} \qquad \text{as } y \to -\infty. \tag{2.14}$$

Since the problem now is axially symmetric, it is natural to make use of cylindrical coordinates $x = r\cos\vartheta$, $z = r\sin\vartheta$, $y = y$. Thus equation (2.12) reads

$$\frac{1}{r}\frac{\partial}{\partial r}\left(r\,\frac{\partial\hat{\varphi}}{\partial r}\right) + \frac{\partial^2\hat{\varphi}}{\partial y^2} = 0. \tag{2.15}$$

Again we introduce dimensionless coordinates:

$$\bar{r} = \frac{r\omega^2}{g}, \qquad \bar{y} = \frac{y\omega^2}{g}.$$

The transform leaves the differential equation (2.15) invariant, but the boundary condition (2.13) becomes

$$\hat{\varphi} = \hat{\varphi}_{\bar{y}} \qquad \text{for } \bar{y} = 0. \tag{2.16}$$

We solve this problem by the method of separation of variables and assume that

$$\hat{\varphi}(\bar{r},\bar{y}) = e^{\lambda \bar{y}} R(\bar{r}),$$

where $R(\bar{r})$ is a solution of the ordinary differential equation

$$\frac{1}{\bar{r}} \frac{d}{d\bar{r}} \left(\bar{r} \frac{dR}{d\bar{r}} \right) + \lambda^2 R = 0.$$

The boundary condition (2.16) gives that $\lambda = 1$. Thus we obtain

$$\hat{\varphi}(\bar{r},\bar{y}) = e^{\bar{y}} \left\{ AH_0^{(1)}(\bar{r}) + BH_0^{(2)}(\bar{r}) \right\}, \tag{2.17}$$

where $H_0^{(i)}$ are Hankel functions of order zero, and A and B are constants to be determined from the radiation condition as follows.

As is well-known, for large value $\bar{r}$ we have

$$H_0^{(1)}(\bar{r}) \sim \sqrt{\frac{2}{\pi \bar{r}}}\, e^{i(\bar{r} - \frac{\pi}{4})}, \text{ and}$$

$$H_0^{(2)}(\bar{r}) \sim \sqrt{\frac{2}{\pi \bar{r}}}\, e^{-i(\bar{r} - \frac{\pi}{4})}.$$

With time dependence $e^{i\omega t}$, only the solution

$$\varphi(\bar{r},\bar{y},\bar{t}) = Be^{\bar{y}} H_0^{(2)}(\bar{r})\, e^{i\omega t} \tag{2.18}$$

represents outgoing waves. For large $\bar{r}$ it behaves as

$$\varphi(\bar{r},\bar{y},\bar{t}) \sim Be^{\bar{y}} \sqrt{\frac{2}{\pi \bar{r}}}\, e^{-i(\bar{r} - \omega t - \frac{\pi}{4})},$$

and the phase is defined by

$$\bar{r} - \omega t = \frac{\omega^2}{g}\left(r - \frac{g}{\omega}t\right),$$

which gives g/ω for the *phase velocity*.

The water height η is given by

$$\eta(\bar{r},t) = -\frac{i\omega}{g} BH_0^{(2)}(\bar{r})\, e^{i\omega t} \tag{2.19}$$

from equations (2.11), (1.33) and (2.18). Here it is understood that either the real or the imaginary part of the right hand side of (2.19) is to be taken. Henceforth the same agreement applies hereafter.
This is an example of centered outgoing waves.

2.3 Travelling Waves

It is easy to obtain travelling plane waves. As in Chapter one for small amplitude waves the linearized problem is defined by (1.32). For simplicity we restrict ourselves to the situation when $U = 0$. We consider two cases according to the water depth. We begin with the infinite depth. In this case the boundary value problem (1.32) consists of the Laplace equation

$$\varphi_{xx} + \varphi_{yy} + \varphi_{zz} = 0 \tag{2.20}$$

together with the surface conditions

$$\varphi_{tt} + g\varphi_y = 0 \qquad \text{at } y = 0 \tag{2.21}$$

and the condition at infinity

$$\varphi_y \to 0 \qquad \text{as } y \to -\infty. \tag{2.22}$$

We seek a solution $\varphi(x,y,z;t)$ of (2.20)-(2.22) in the form

$$\varphi(x,y,z;t) = Ae^{i(\alpha x + \beta z) + ky + i\omega t}, \tag{2.23}$$

where α, β, k, ω and A are constants. Clearly (2.22) will be satisfied if k is positive. Substituting (2.23) into (2.20) and (2.21) we obtain

$$\begin{aligned} k^2 &= \alpha^2 + \beta^2 \quad \text{and} \\ -\omega^2 + gk &= 0. \end{aligned} \tag{2.24}$$

Set $\alpha = -k \cos \vartheta$ and $\beta = -k \sin \vartheta$ which clearly satisfy the first equation of (2.24) for any k. The second one gives that $k = (\omega^2/g)$ which is known as the *dispersion relation* - a relation between wave number k and frequency ω. Then the potential function takes the form

$$\varphi(x,y,z;t) = A \exp\left\{-i\omega\left[\frac{\omega}{g}(x \cos\vartheta + z \sin\vartheta) - t\right] + \frac{\omega^2}{g} y\right\}, \tag{2.25}$$

and consequently the water height is given by

$$\begin{aligned} \eta(x,z;t) &= -\frac{1}{g}\varphi_t \\ &= -A\frac{i\omega}{g} \exp\left\{-i\omega\left[\frac{\omega}{g}(x\cos\vartheta + z\sin\vartheta) - t\right]\right\} \end{aligned} \tag{2.26}$$

through use of (1.33). This formula represents plane waves. Observe that the preceding result (2.19) can be obtained by integrating (for complex ϑ) these plane waves.

For $\vartheta = 0$, we have plane waves travelling along the x-axis:

$$\begin{aligned} \eta(x,o;t) &= -\frac{i\omega}{g} A e^{-i\left(\frac{\omega^2}{g} x - \omega t\right)} \\ &= A_1(\omega)\, e^{-i \frac{\omega^2}{g}(x - ct)}, \end{aligned}$$

where $c = g/\omega$ is the velocity of the wave (or *phase velocity*) and $A_1(\omega) = -(i\omega/g)A$.

The corresponding problems for finite depth can be treated in the same way. We write

$$\varphi(x,y,z) = \hat{\varphi}(x,y,z)\, e^{i\omega t}.$$

Then in this case we have from (2.21) the surface condition

$$\hat{\varphi}_y = \frac{\omega^2}{g}\hat{\varphi} \qquad \text{at } y = 0, \tag{2.27}$$

while the condition at infinity (2.22) is replaced by the boundary condition (1.32). In terms of $\hat{\varphi}$, we have

$$\hat{\varphi}_y = 0 \qquad \text{at } y = -h. \tag{2.28}$$

For the travelling plane wave in the direction of the x-axis, i.e., $\hat{\varphi} = \hat{\varphi}(x,y)$, a simple manipulation by the method of separation of variables leads to the solution

$$\varphi(x,y,t) = A \cosh k(y + h)\, e^{-i(kx - \omega t)}, \tag{2.29}$$

where the wave number k and frequency ω are related by the dispersion relation

$$\omega^2 = gk \tanh kh. \tag{2.30}$$

Waves with different wave number travel with different phase velocity c which is defined by

$$c = \frac{\omega}{k} = \sqrt{\frac{g \tanh kh}{k}}. \tag{2.31}$$

Note that for kh small, since $\tanh kh = kh + O((kh)^3)$, we have $c = \sqrt{gh}$ which is the case of no dispersion. Observe again that if we let $h \to \infty$, we recover the case of infinite depth, (2.24). The dispersion causes a wave pattern, which at a certain place x and time t is a superposition of harmonic waves to be distorted at other places, because the components travel with different velocities. In the case of dispersion, it is difficult to determine the concept of 'wave speed'.

2.4 Group Velocity

For the description of plane waves it is sufficient to restrict the considerations to the one dimensional case. We represent at $t = 0$ the water height $\eta(x,o)$ by a real integral

$$\eta(x) = \int_0^\infty C(k) \cos kx \, dk + \int_0^\infty S(k) \sin kx \, dk \qquad (2.32)$$

with

$$C(k) = \frac{1}{\pi} \int_0^\infty \eta(x) \cos kx \, dx, \text{ and}$$

$$S(k) = \frac{1}{\pi} \int_0^\infty \eta(x) \sin kx \, dx.$$

Since $C(k)$ and $S(k)$ are respectively even and odd functions, setting

$$A(k) = \frac{1}{2} \Big(C(k) + iS(k) \Big), \qquad (2.33)$$

we can rewrite $\eta(x)$ as a complex integral

$$\eta(x) = \int_{-\infty}^\infty A(k) \, e^{-ikx} \, dk. \qquad (2.34)$$

A simple calculation shows that

$$\eta(x) = 2\mathrm{Re} \int_0^\infty A(k) \, e^{-ikx} \, dk$$

$$= \int_{-\infty}^\infty A^*(k) \, e^{ikx} \, dk, \qquad (2.35)$$

where $A^*(k)$ is the complex conjugate of $A(k)$.

For an understanding of the wave dispersion phenomenon, it is necessary to consider the energy propagation in the wave (linearized approximation). If the function $\eta(x)$ belongs to L^2, i.e. $\int_{-\infty}^\infty \eta(x)^2 \, dx$ exists, the potential energy in the wave is given by

$$E = \frac{1}{2}\rho g \int_{-\infty}^{\infty} \eta(x)^2\, dx$$

$$= \frac{1}{2}\rho g \int_{-\infty}^{\infty} \left(\int_{-\infty}^{\infty} A(k)\, e^{-ikx}\, dk\right) \times \left(\int_{-\infty}^{\infty} A^*(k')\, e^{ik'x}\, dk'\right) dx$$

from (2.34) and (2.35). The latter integral can now be calculated by making use of the Fourier inversion theorem and the fact that $\int_{-\infty}^{\infty} e^{i(k' - k)x}\, dx = 2\pi\, \delta(k' - k)$.
This gives

$$E = \frac{1}{2}\rho g\, 2\pi \int_{-\infty}^{\infty} |A(k)|^2\, dk. \tag{2.36}$$

Hence from (2.33) we have

$$E = \rho g \pi \int_{-\infty}^{\infty} \left\{C(k)^2 + S(k)^2\right\} dk. \tag{2.37}$$

If the dispersion relation $\omega = \omega(k)$ is known (for convenience we extend the definition of $\omega(k)$ to negative k by setting $\omega(-k) = -\omega(k)$), then we can compute the water height η at any arbitrary time t as follows:

$$\eta(x,t) = \int_{-\infty}^{\infty} A(k)\, e^{i(\omega t - kx)}\, dk$$

$$= \int_{-\infty}^{\infty} A(k)\, e^{-ik(x - ct)}\, dk \tag{2.38}$$

in terms of the phase velocity $c = \omega/k$. Here it is assumed, that the initial conditions are such that the wave propagates only in the projection of the positive x-axis.

The total potential energy is conserved; the wave only changes the distribution of the energy along the x-axis. In fact we have

$$E(t) = \frac{1}{2} \rho g \int_{-\infty}^{\infty} |\eta(x,t)|^2 \, dx$$

$$= \frac{1}{2} \rho g \int_{-\infty}^{\infty} dx \int_{-\infty}^{\infty} A(k) \; e^{i(\omega t - kx)} \, dk$$

$$\times \int_{-\infty}^{\infty} A^*(k') \; e^{-i(\omega' t - k'x)} \, dk'$$

with $\omega' = \omega(k')$. This latter integral follows from (2.38) and can be calculated similarly according to the Fourier inversion theorem. We find again

$$E(t) = \rho g \pi \int_{-\infty}^{\infty} |A(k)|^2 \, dk. \tag{2.39}$$

Now we are going to find a measure for the velocity of the energy propagation and calculate to this end the location of the 'center of gravity' $\bar{x}(t)$ of the first moment of the energy, which is defined by

$$\bar{x}(t) = \frac{\int_{-\infty}^{\infty} x|\eta(x,t)|^2 \, dx}{\int_{-\infty}^{\infty} |\eta(x,t)|^2 \, dx}, \tag{2.40}$$

provided both integrals exist. Here the denominator has been shown to be a constant in time and can be calculated easily from (2.39). The numerator, however, requires some investigation. Equation (2.38) yields

$$\int_{-\infty}^{\infty} x|\eta(x,t)|^2 \, dx = \int_{-\infty}^{\infty} x \, dx \int_{-\infty}^{\infty} A(k) \; e^{i(\omega t - kx)} \, dk$$

$$\times \int_{-\infty}^{\infty} A^*(k') \; e^{-i(\omega' t - k'x)} \, dk'$$

$$= i \int_{-\infty}^{\infty} dx \int_{-\infty}^{\infty} A(k) d \left(e^{i(\omega t - kx)} \right)$$

$$\times \int_{-\infty}^{\infty} A^*(k') \; e^{-i(\omega' t - k'x)} \, dk'$$

$$+ \; t \int_{-\infty}^{\infty} dx \int_{-\infty}^{\infty} A(k) \frac{d\omega(k)}{dk}$$

$$\times \int_{-\infty}^{\infty} A^*(k') \; e^{i(\omega t - kx) - i(\omega' t - k'x)} \, dk \, dk'$$

$$:= I_1 + I_2.$$

Integrating by parts and taking account of the fact that $A(k) \to 0$, as $k \to \pm\infty$ in view of Bessel's inequality, we find

$$I_1 = -i \int_{-\infty}^{\infty} dx \int_{-\infty}^{\infty} \frac{dA(k)}{dk}$$

$$\times \int_{-\infty}^{\infty} A^*(k') \; e^{i(\omega t - kx) - i(\omega' t - k'x)} \, dk \, dk'.$$

Then by the Fourier inversion formula, we obtain:

$$I_1 = -2\pi i \int_{-\infty}^{\infty} \frac{dA(k)}{dk} A^*(k) \, dk;$$

$$I_2 = 2\pi t \int_{-\infty}^{\infty} \frac{d\omega(k)}{dk} \; |A(k)|^2 \, dk.$$

Adding I_1 and I_2, we have

$$\int_{-\infty}^{\infty} x|\eta(x,t)|^2 \, dx = 2\pi \left\{ -i \int_{-\infty}^{\infty} \frac{dA(k)}{dk} A^*(k) \, dk + t \int_{-\infty}^{\infty} \frac{d\omega(k)}{dk} \; |A(k)|^2 \, dk \right\}. \tag{2.41}$$

We define, as a mean value of a quantity $\psi(k)$ in the k-domain,

$$\bar{\Psi} = \frac{\int_{-\infty}^{\infty} \psi(k) \; |A(k)|^2 \, dk}{\int_{-\infty}^{\infty} |A(k)|^2 \, dk}, \tag{2.42}$$

and remark that the first term in (2.41) determines the position of $\bar{x}(t)$ for $t = 0$. Hence we find

$$\bar{x}(t) = \bar{x}(0) + t \, \overline{\frac{d\omega}{dk}}, \tag{2.43}$$

i.e. the center of gravity propagates with a velocity which is equal to the mean value of $\frac{d\omega(k)}{dk}$. Here $\frac{d\omega(k)}{dk}$ is called the *group velocity*; hence the mean value of the group velocity $\overline{d\omega}/dk$ is a measure for the speed of propagation of the energy.

The significance of this result becomes clear when we consider an amplitude spectrum $A(k)$, which extends only over a narrow wave number band:

$$\eta(x,t) = \int_{k_0-\varepsilon}^{k_0+\varepsilon} A(k)\, e^{-i(kx-\omega(k)t)}\, dk \qquad (2.44)$$

$$\varepsilon > 0.$$

The center of gravity satisfies

$$\bar{x}(t;k_0,\varepsilon) = \bar{x}(o;k_0,\varepsilon) + t\, \overline{\omega'(k_0,\varepsilon)}, \qquad (2.45)$$

where $\bar{\omega}'(k_0,\varepsilon)$ is now the mean value of $\omega'(k) = \frac{d\omega}{dk}$ over the narrow band $[k_0-\varepsilon,\ k_0+\varepsilon]$. For small values of ε we simply replace $\overline{\omega'(k_0,\varepsilon)}$ by $\omega'(k_0)$.

For small values of t, one can make a more accurate analysis of the motion as follows. Expanding $\omega(k)$ in the form

$$\omega(k) = \omega(k_0) + (k-k_0)\,\omega'(k_0) + \frac{(k-k_0)^2}{2}\,\tilde{\omega}''(k),$$

and substituting into (2.44), we may then write

$$\eta(x,t) = \int_{k_0-\varepsilon}^{k_0+\varepsilon} A(k) \exp\Big\{-i(k_0x-\omega(k_0)t) - i(k-k_0)$$

$$\times (x-\omega'(k_0)t) + \frac{i(k-k_0)^2}{2}\,\tilde{\omega}''(k)t\Big\}\, dk$$

$$= e^{i(k_0x-\omega(k_0)t)} \int_{k_0-\varepsilon}^{k_0+\varepsilon} A(k)\, e^{-i(k-k_0)(x-\omega'(k_0)t)}$$

$$\times\, dk + R, \qquad (2.46)$$

where

$$R = e^{-i(k_0x-\omega(k_0)t)} \int_{k_0-\varepsilon}^{k_0+\varepsilon} A(k)\ e^{-i(x-\omega'(k_0)t)(k-k_0)}$$

$$\times \left\{\exp\left[\frac{i(k-k_0)^2}{2}\ \tilde{\omega}''(k)t\right] - 1\right\}.$$

Using the inequality $|e^{iu} - 1| \leqslant |u|$, we find an estimate of the remainder

$$|R| \leqslant \int_{k_0-\varepsilon}^{k_0+\varepsilon} |A(k)| \left|\frac{(k-k_0)^2}{2}\ \tilde{\omega}''(k)t\right|\ dk$$

$$\leqslant \frac{1}{3}\left(\max_{|k-k_0|<\varepsilon} |A(k)|\right)\left(\max_{|k-k_0|<\varepsilon} |\omega''(k)|\right) \varepsilon^3 t,$$

which shows that for not too large values of t, the first term of (2.46) gives a good approximation of η. Assuming, for small ε, A(k) to be a constant $A(k_0)$ over the interval, we can integrate:

$$A(k_0)\ e^{-i(k_0x-\omega(k_0)t)} \int_{k_0-\varepsilon}^{k_0+\varepsilon} e^{-i(k-k_0)(x-\omega'(k_0)t)}\ dk$$

$$= A(k_0)\ e^{-\frac{i}{k_0}\left(x - \frac{\omega(k_0)}{k_0}\ t\right)}\ \frac{2\sin[(x-\omega'(k_0)t)\varepsilon]}{x - \omega'(k_0)t}.$$

Hence we have, for small ε and t not too large

$$\eta(x,t) \cong A(k_0)\ e^{-\frac{i}{k_0}\left(x - \frac{\omega(k_0)}{k_0}\ t\right)}\ \frac{2\sin[(x-\omega'(k_0)t)\varepsilon]}{x - \omega'(k_0)t} \tag{2.47}$$

as shown in Figure 2.2.

This represents a modulated wave; the amplitude moves with the group velocity $\omega'(k_0)$ (the dotted enveloping curves) while the phase moves with phase velocity $\omega(k_0)/k_0$ (the inscribed solid curves).

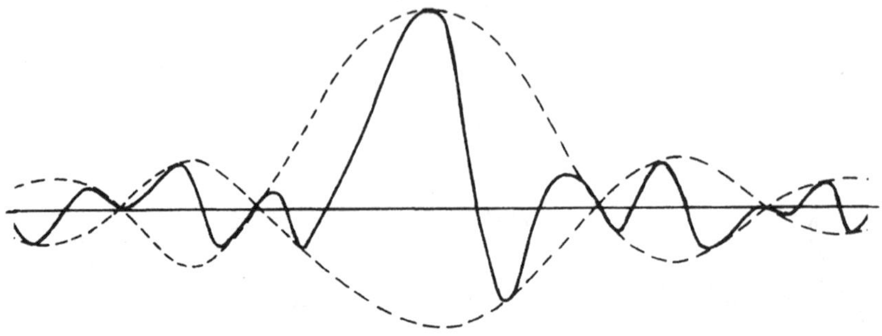

Figure 2.2

2.5 Wave Fronts

In view of (2.26) and (2.38), we now consider the general representation for the free surface elevation:

$$\eta(x,z,t) = \int_{-\infty}^{\infty} A(k)\, e^{-i(kx\cos\vartheta + kz\sin\vartheta - \omega t)}\, dk. \tag{2.48}$$

In particular, we are interested in the asymptotic behaviour of η for large values of t. We apply the method of stationary phase to (2.48). The method requires the determination of the value k for which the phase

$$\begin{aligned}\psi(x,z,k) &= -\omega t + k(x\cos\vartheta + z\sin\vartheta)\\ &= -t\left[k\left(\frac{x}{t}\cos\vartheta + \frac{z}{t}\sin\vartheta\right) - \omega(k)\right]\end{aligned} \tag{2.49}$$

is stationary. (Here we consider ψ as depending upon the three parameters $\frac{x}{t}$, $\frac{z}{t}$ and t. For each pair of values of $\frac{x}{t}$ and $\frac{z}{t}$, the asymptotic expansion for η is considered for large t). This leads to the consideration of solutions of the equations

$$\frac{d\psi}{dk} = 0 \quad \text{or} \quad -\frac{d\omega}{dk}t + x\cos\vartheta + z\sin\vartheta = 0. \tag{2.50}$$

Let k_0 be any solution of (2.50). Then the approximate result for large t is

$$\eta(x,z,t) = A(k_0)\sqrt{\frac{2\pi}{t|\omega''(k_0)|}}$$
$$\times\, e^{-i(k_0 x\cos\vartheta + k_0 z\sin\vartheta - \omega(k_0)t - \frac{\pi}{4}\operatorname{sgn}\omega''(k_0))} \tag{2.51}$$

provided that $\dfrac{d^2\psi(k_0)}{dk^2} \neq 0$, i.e. $\omega''(k_0) \neq 0$.

The lines ψ = constant are the lines of constant phase; these lines are called *wave fronts*. We can find a partial differential equation for the wave fronts from the dispersion relation $\omega = H(k)$. In fact, we can express k_0 in terms of x,z,t and ϑ from (2.50) so that differentiations of (2.49) (with $k = k_0$) with respect to these variables yield

$$\psi_x = k_0\cos\vartheta + (x\cos\vartheta + z\sin\vartheta - \omega_0't)\frac{\partial k_0}{\partial x} = k_0\cos\vartheta,$$

$$\psi_z = k_0\sin\vartheta + (x\cos\vartheta + z\sin\vartheta - \omega_0't)\frac{\partial k}{\partial z} = k_0\sin\vartheta,$$

$$\psi_t = -\omega_0 + (x\cos\vartheta + z\sin\vartheta - \omega't)\frac{\partial k_0}{\partial t} = -\omega_0 \tag{2.52}$$

with $\omega_0 = H(k_0)$. The first two equations of (2.52) imply that

$$k_0^2 = \psi_x^2 + \psi_z^2$$

with which the third one shows that the dispersion relation $\omega_0 = H(k_0)$ gives a partial differential equation for the phase function ψ, the Hamilton-Jacobi equation

$$\psi_t + H\left(\sqrt{\psi_x^2 + \psi_z^2}\right) = 0$$

or

$$\psi_t + H\left(\sqrt{p^2 + q^2}\right) = 0, \tag{2.53}$$

where $p = \psi_x = k_0\cos\vartheta$ and $q = \psi_z = k_0\sin\vartheta$ are the conjugate variables to x and z, respectively. We have just seen that the wave fronts correspond to level curves of the Hamilton-Jacobi equation. But in wave phenomena one expects the dual concept of rays to appear also. The rays in the present case are the *characteristics* of the above Hamilton-Jacobi equation, i.e. the solutions of the system of ODE:

$$\frac{dx}{dt} = \frac{\partial H}{\partial p}, \quad \frac{dp}{dt} = -\frac{\partial H}{\partial x} = 0,$$
$$\frac{dz}{dt} = \frac{\partial H}{\partial q}, \quad \frac{dq}{dt} = -\frac{\partial H}{\partial z} = 0. \tag{2.54}$$

(see (A.2) for a brief summary of the concepts of characteristics). From the equations (2.54) it is easy to see that in the x,z,t-space, the characteristics are straight lines for constant t as in (2.50).

2.6 Wave Patterns

In §2.5, the Hamilton-Jacobi equation (2.53) for the wave fronts was derived from the equations in a rather complicated way. At first we gave an exact solution η of the linearized problem (defined by (1.32), (1.33) with $U = 0$, to which we later applied an asymptotic expansion, which resulted in a first order partial differential equation. The result obtained is more or less similar to the characteristic equation for hyperbolic equations, although the wave fronts are by no means characteristic surfaces for the equations, which do not even have real characteristics.

In order to give a direct derivation we first define a *wave front* on the two dimensional x,y-plane as a curve along which a transverse derivative of a solution φ of the equation considered is much larger than the tangential derivative. This means that, introducing new coordinates ξ_1 transverse to the wave fronts and ξ_2 along the wave fronts (Figure 2.3), we must have that $\varphi_{\xi_1} \gg \varphi_{\xi_2}$, i.e. there should exist a constant $K \gg 1$ such that $\varphi_{\xi_1} \approx K\varphi_{\xi_2}$. Here ξ_1 and ξ_2 are supposed to be functions of x and y with derivatives of order unity with respect to K. We introduce a new coordinate $s = K\xi_1$ such that

$$\varphi_s = \frac{1}{K}\varphi_{\xi_1} = O(1). \tag{2.55}$$

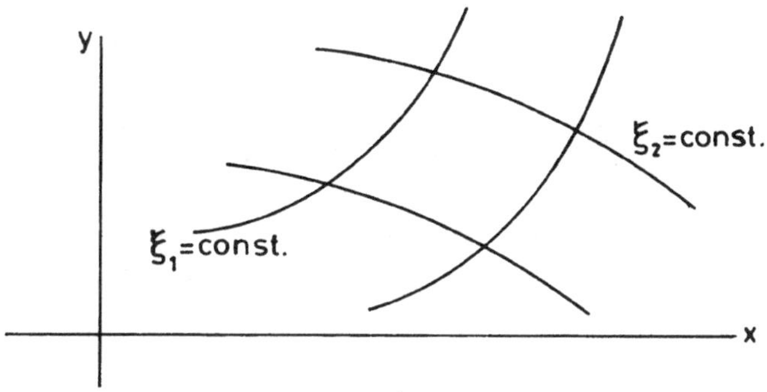

Figure 2.3

We now illustrate this procedure by considering a simpler equation than the equation of water waves, the Klein-Gordon equation in dimensionless form:

$$\varphi_{xx} - \varphi_{tt} - a^2\varphi = 0, \tag{2.56}$$

where a is a constant. We first derive the Hamilton-Jacobi equations for the phase function to the method used in §2.5 and we will refer to it as an indirect method. For solutions of the form $Ae^{i(kx-\omega t)}$ we easily find the dispersion relation between k and ω

$$\omega = \sqrt{a^2 + k^2} := H(k), \tag{2.57}$$

which gives the Hamilton-Jacobi equation from (2.53) with $\psi = J$:

$$J_t + H(J_x) = 0, \tag{2.58}$$

where $J_x = k$. The characteristics of (2.58) are solutions of the equations:

$$\begin{aligned} \frac{dx}{dt} &= \frac{\partial H}{\partial J_x} = \frac{k}{\sqrt{a^2 + k^2}}, \\ \frac{dp}{dt} &= \frac{dJ_x}{dt} = -\frac{\partial H}{\partial x} = 0. \end{aligned} \tag{2.59}$$

Thus the characteristics are the straight lines of the form

$$x - \frac{k}{\sqrt{a^2 + k^2}} t = \text{constant}, \tag{2.60}$$

corresponding to the group velocity $\frac{dH}{dk} = \frac{k}{\sqrt{a^2 + k^2}}$.

Now let us examine the above problem by the direct method. Using equation (2.55), a straightforward calculation shows that

$$\begin{aligned} \varphi_{xx} &= K^2\varphi_{ss}\xi_{1x}^2 + K\left(2\varphi_{s\xi_2}\xi_{1x}\xi_{2x} + \xi_{1xx}\varphi_s\right) \\ &\quad + \varphi_{\xi_2\xi_2}\xi_{2x}^2 + \varphi_{\xi_2}\xi_{2xx}, \\ \varphi_{tt} &= K^2\varphi_{ss}\xi_{1t}^2 + K\left(2\varphi_{s\xi_2}\xi_{1t}\xi_{2x} + \xi_{1xx}\varphi_s\right) \\ &\quad + \varphi_{\xi_2\xi_2}\xi_{2t}^2 + \varphi_{\xi_2}\xi_{2tt}. \end{aligned} \tag{2.61}$$

Substituting into (2.56) gives

$$\begin{aligned} &K^2\varphi_{ss}\left(\xi_{1x}^2 - \xi_{1t}^2\right) + K\varphi_s\left(\xi_{1xx} - \xi_{1tt}\right) + \\ &\quad + 2K\varphi_{s\xi_2}\left(\xi_{1x}\xi_{2x} - \xi_{1t}\xi_{2t}\right) + \varphi_{\xi_2\xi_2}\left(\xi_{2x}^2 - \xi_{2t}^2\right) \\ &\quad + \varphi_{\xi_2}\left(\xi_{2t} - \xi_{2tt}\right) - a^2\varphi = 0. \end{aligned}$$

As $K \to \infty$, we obtain the characteristic equation for (2.56). This is obvious because a characteristic would be a line along which the second order derivatives may be discontinuous. Now, regarding the constant a as a large number with respect to some reference length and identifying K with a, we have the equation

$$\varphi_{ss}\left(\xi_{1x}^2 - \xi_{1t}^2\right) - \varphi = 0. \tag{2.63}$$

to a first order of approximation. If we want this equation to represent the motion along the wave fronts, we must put the term $(\xi_{1x}^2 - \xi_{1t}^2)$ equal to a constant which we choose to be -1, i.e.,

$$\xi_{1x}^2 - \xi_{1t}^2 = -1. \tag{2.64}$$

Clearly, this gives immediately the Hamilton-Jacobi equation, $\xi_{1t} = \sqrt{1 + \xi_{1x}^2}$, which reduces to the equation (2.58) with ξ_1 replaced by $(-1/a)J$.

The same scheme can be applied to the problem of the moving singularity defined by the time independent form (1.32) and (1.33), i.e.:

$$\varphi_{xx} + \varphi_{yy} + \varphi_{zz} = 0,$$

$$U^2\varphi_{xx} + g\varphi_y = 0, \quad \text{for } y = 0.$$

In terms of the dimensionless variables $\bar{x} = \frac{x}{L}$, $\bar{y} = \frac{y}{L}$ and $\bar{z} = \frac{z}{L}$, we have

$$\begin{aligned} &\varphi_{\bar{x}\bar{x}} + \varphi_{\bar{y}\bar{y}} + \varphi_{\bar{z}\bar{z}} = 0, \\ &\varphi_{\bar{x}\bar{x}} + \left(\frac{gL}{U^2}\right)\varphi_{\bar{y}} = 0, \quad \text{for } \bar{y} = 0. \end{aligned} \tag{2.65}$$

Here L denotes a proper reference length.

We are only interested in the *wave pattern*, hence in the lines of constant phase of η (which from (1.33) amounts to the same as for $\varphi_{\bar{x}}$ at $\bar{y} = 0$). We further remark that from the nature of the equations (2.65) we know that the wave is only appreciable at the upper layer of the water. Hence when we introduce the coordinates ξ_1 and ξ_2 on the x,z-plane, where the lines ξ_1 = constant represent wave fronts, the derivative φ_{ξ_1} is large with respect to φ_{ξ_2} but the derivative $\varphi_{\bar{y}}$ must be of the same order of magnitude as φ_{ξ_1}. Therefore, we introduce a coordinate $s = K\xi_1$ and a coordinate $Y = K\bar{y}$ in terms of which we have

$$\varphi_{\bar{x}\bar{x}} = K^2\varphi_{ss}\xi_{1\bar{x}}^2 + K\left(2\varphi_{s\xi_2}\xi_{1\bar{x}}\xi_{2\bar{x}} + \xi_{1\bar{x}\bar{x}}\varphi_s\right) + \varphi_{\xi_2\xi_2}\xi_{2\bar{x}}^2 + \varphi_{\xi_2}\xi_{2\bar{x}\bar{x}},$$

$$\varphi_{\bar{z}\bar{z}} = K^2\varphi_{ss}\xi_{1\bar{z}}^2 + K\left(2\varphi_{s\xi_2}\xi_{1\bar{z}}\xi_{2\bar{z}} + \xi_{1\bar{z}\bar{z}}\varphi_s\right) + \varphi_{\xi_2\xi_2}\xi_{2\bar{z}}^2 + \varphi_{\xi_2}\xi_{2\bar{z}\bar{z}},$$

and

$$\varphi_{\bar{Y}\bar{Y}} = K^2 \varphi_{YY}.$$

From equations (2.65), we then have

$$K^2 \varphi_{ss}\left(\xi^2_{1\bar{x}} + \xi^2_{1\bar{z}}\right) + K^2 \varphi_{YY} + 0(K) = 0, \qquad (2.66)$$

together with the surface condition

$$K^2 \varphi_{ss} \xi^2_{1\bar{x}} + K\left(\frac{gL}{U^2}\right) \varphi_Y + 0(K) = 0,$$

$$\text{for } Y = 0. \qquad (2.67)$$

This yields the first approximation:

$$\varphi_{ss}\left(\xi^2_{1\bar{x}} + \xi^2_{1\bar{z}}\right) + \varphi_{YY} = 0,$$

$$\varphi_{ss}\xi^2_{1\bar{x}} + \varphi_Y = 0, \qquad \text{for } Y = 0, \qquad (2.68)$$

where we identify K with gL/U^2.

Since $\xi^2_{1\bar{x}} + \xi^2_{1\bar{z}}$ and $\xi^2_{1\bar{x}}$ are slowly varying variables, we introduce constants, α and β defined by

$$\alpha^2 = \xi^2_{1\bar{x}} + \xi^2_{1\bar{z}}, \text{ and } \beta = \xi^2_{1\bar{x}}.$$

This leads to the problem:

$$\alpha^2 \varphi_{ss} + \varphi_{YY} = 0,$$

$$\beta\varphi_{ss} + \varphi_y = 0, \qquad \text{for } Y = 0,$$

which has a solution

$$\varphi = e^{is+\alpha Y}.$$

This solution which goes to zero as $Y \to -\infty$ $(\alpha > 0)$ can satisfy the surface condition only if

$$\alpha = \beta$$

or

$$\xi^2_{1\bar{x}} + \xi^2_{1\bar{z}} = \xi^4_{1\bar{x}}. \qquad (2.69)$$

It should be emphasized that these considerations are only valid to an order of approximation of 1/K. The present approach is a variation of the ray method in geometrical optics. Higher order approximations can be derived in a similar manner.

The characteristic equations of the partial differential equation (2.69) take the form:

$$\dot{\bar{x}} = 4p^3 - 2p; \qquad \dot{p} = 0,$$

$$\dot{\bar{z}} = -2q; \qquad \dot{q} = 0,$$

$$\dot{\xi}_1 = (4p^4 - 2p^2 - 2q^2),$$

with $p = \xi_{1\bar{x}}$ and $q = \xi_{1\bar{z}}$, where the dot . notation denotes differentiation with respect to some parameter, say, τ. Hence p and q are constants and we have the parametric equations for the rays:

$$\begin{aligned} \bar{x} &= 2p(2p^2 - 1)\,\tau, \\ \bar{z} &= -2q\,\tau, \\ \xi_1 &= (4p^4 - 2p^2 - 2q^2)\,\tau. \end{aligned} \tag{2.70}$$

From equation (2.69) we have

$$q = -p\sqrt{p^2 - 1}. \tag{2.71}$$

To eliminate τ from (2.70) by making use of (2.71), we finally obtain

$$\bar{x} = \xi_1 \frac{(2p^2 - 1)}{p^3}, \qquad \bar{z} = \frac{\xi_1 \sqrt{p^2 - 1}}{p^3},$$

which reduces to equations (2.10) if we set $p = -\frac{1}{\cos \vartheta}$. This shows that the curves ξ_1 = constant are indeed the curves of constant phase.

CHAPTER III - IRREGULAR AND NON-LINEAR WAVES

The surface waves of the sea are almost always random in the sense that detailed configuration of the surface varies in an irregular manner in both space and time. Section 3.1 contains a brief description of the Wiener spectrum in connection with generalized Fourier representations for the surface waves [2], [13]. In this way we see how one may represent the surface wave by a superposition of harmonic waves with amplitudes being a stochastic process.

The remaining sections in the chapter are devoted to the non-linear waves. In Section 3.2, we give a systematic derivation of the shallow water theory from the exact hydrodynamical equations as the approximation of lowest order in a perturbation procedure. Here the relevant small parameter is the ratio of the depth of water to some other characteristic length associated with the horizontal direction such as the wave length; the water is considered to be shallow when this parameter is small. It is a different kind of approximation from the previous linear theory for waves of small amplitude. The resulting equations here are quasi-linear and are exactly analogous to the ones in gas dynamics. Second order approximations are included in the last Section 3.3. In particular, an asymptotic theory will be developed for slowly varying wave trains, which may be considered as nearly uniform in the regions of order of magnitude of a small number of wave lengths and periods. Some non-linear dispersive wave phenomena will be discussed and more details can be found in [3].

3.1 The Wiener Spectrum

The actual motion of the sea is by no means a harmonic motion with constant frequency. In §2.4, we gave a representation of the sea surface $\eta(x,t)$ in the form of an integral

$$\eta(x,t) = \int_{-\infty}^{\infty} \left\{ A(k)\ e^{-i(kx-\omega t)} + A^*(k)\ e^{i(kx-\omega t)} \right\} dk$$

under the hypothesis that $\int_{-\infty}^{\infty} |A(k)|^2\, dk$ is convergent. It is, however, not appropriate to suppose that the amplitudes of the high frequencies go to zero. The representation by a Fourier integral there breaks down and we must use a theory developed by Wiener of which we shall give an outline below. Details of proofs can be found in Wiener's book in the References.

For a better understanding, we begin with a trigonometric polynomial f(t) defined by

$$f(t) = \sum_{j=1}^{n} A_j e^{i\lambda_j t},$$

and calculate

$$f(t+\tau) f^*(\tau) = \sum_{j=1}^{n} \sum_{k=1}^{n} A_j A_k^*\ e^{i\lambda_j t}\ e^{i(\lambda_j - \lambda_k)\tau}.$$

Here A_k^* again denotes the complex conjugate of A_k. Now by taking the averaged mean of this quantity, we obtain that

$$\lim_{T\to\infty} \frac{1}{2T} \int_{-T}^{T} f(t+\tau) f^*(\tau)\ d\tau$$

$$= \sum_{j=1}^{n} \sum_{k=1}^{n} A_j A_k^*\ e^{i\lambda_j t} \lim_{T\to\infty} \frac{1}{2T} \int_{-T}^{T} e^{i(\lambda_j - \lambda_k)\tau}\ d\tau.$$

For $\lambda_j = \lambda_k$, the limit is equal to unity, but for $\lambda_j \neq \lambda_k$, the limit vanishes. This leads to the result

$$\lim_{T\to\infty} \frac{1}{2T} \int_{-T}^{T} f(t+\tau) f^*(\tau) d\tau = \sum_{j=1}^{n} |A_j|^2\ e^{i\lambda_j t}.$$

In other words, if we put

$$\varphi(t) = \lim_{T\to\infty} \frac{1}{2T} \int_{-T}^{T} f(t + \tau) f^*(\tau)\, d\tau,$$

then $\varphi(t)$ exists for every t, is continuous and consists of terms with the same frequency as those constituting f(t) and with amplitudes equal to the square of the amplitudes of the corresponding terms of f(t). Hence we see a way of finding a Fourier integral representation of a function which does not vanish at infinity. One can analyze the surface waves by means of functions of this kind in space variables as well as in the time variable.

We next build up the Wiener theory for those complex functions f(t) of the real variable t such that

$$\varphi(t) = \lim_{T\to\infty} \frac{1}{2T} \int_{-T}^{T} f(t + \tau) f^*(\tau)\, d\tau \qquad (3.1)$$

exists for all t. Clearly then $\varphi(0)$ also exists. This means that the quadratic mean of the function exists. Moreover, from the Schwartz's inequality, it can be shown that for all real t:

$$|\varphi(t)| \leqslant \varphi(0).$$

Indeed, it is easy to see that

$$\begin{aligned}
|\varphi(t)| &= \overline{\lim_{T\to\infty}} \left| \frac{1}{2T} \int_{-T}^{T} f(t + \tau) f^*(\tau)\, d\tau \right| \\
&\leqslant \left\{ \overline{\lim_{T\to\infty}} \frac{1}{2T} \int_{-T-|t|}^{T+|t|} |f(\tau)|^2 d\tau \; \overline{\lim_{T\to\infty}} \frac{1}{2T} \int_{-T}^{T} |f(\tau)|^2 d\tau \right\}^{1/2} \\
&\leqslant \left\{ \left(\lim_{T\to\infty} \frac{T+|t|}{T} \right) \left(\overline{\lim_{T\to\infty}} \frac{1}{2(T+|t|)} \int_{-T-|t|}^{T+|t|} |f(\tau)|^2 d\tau \right) \right. \\
&\quad \times \left. \left(\overline{\lim_{T\to\infty}} \frac{1}{2T} \int_{-T}^{T} |f(\tau)|^2 d\tau \right) \right\}^{1/2} \\
&= \left\{ 1 \times \varphi(0) \times \varphi(0) \right\}^{1/2} = \varphi(0).
\end{aligned}$$

Here we use the notation $\overline{\lim}$ for the *limit superior*. Recall that $\overline{\lim}\, S_n = A \Leftrightarrow$ the sequence $\{S_n\}_{n=1}^{\infty}$ is bounded above and has A as its largest limit point.

We now try to define a reasonable Fourier transform for those functions $f(\tau)$ in (3.1). Clearly from (3.1), the standard Fourier transform of $f(\tau)$, $\frac{1}{\sqrt{2\pi}} \int_{-\infty}^{\infty} f(\tau) e^{-i\omega\tau} d\tau$, need not exist but the limit

$$\lim_{T\to\infty} \frac{1}{2T} \int_{-T}^{T} f(\tau)\, e^{i\omega\tau}\, d\tau$$

does exist. This leads us to consider the integral

$$\int_{-\infty}^{\infty} f(t)\, \frac{e^{i\omega t}}{it}\, dt,$$

although the integrand is singular at $t = 0$. Since $\varphi(0)$ exists it follows that $\frac{1}{2T} \int_{-T}^{T} |f(t)|^2\, dt$ is bounded in T and hence the function which is $f(t)/it$ for $|t| > 1$ and 0 for $|t| \leqslant 1$ belongs to L_2; consequently its Fourier transform (in L_2) exists and belongs to L_2. Thus we define

$$S(\omega) = \underset{A\to\infty}{\ell.\text{i.m.}} \frac{1}{\sqrt{2\pi}} \left[\int_{-A}^{-1} + \int_{1}^{A}\right] \frac{f(t)e^{-i\omega t}}{-it}\, dt$$

$$+ \frac{1}{\sqrt{2\pi}} \int_{-1}^{1} f(t)\, \frac{e^{-i\omega t} - 1}{-it}\, dt, \qquad (3.2)$$

and consider the difference $S(\omega+\varepsilon) - S(\omega-\varepsilon)$ which can be reduced to

$$S(\omega+\varepsilon) - S(\omega-\varepsilon) = \underset{A\to\infty}{\ell.\text{i.m.}} \frac{1}{\sqrt{2\pi}} \int_{-A}^{A} f(t)\, \frac{2 \sin \varepsilon t}{t}\, e^{-i\omega t}\, dt. \qquad (3.3)$$

Here and in the sequel we use the notation that $\underset{\alpha\to\alpha_0}{\ell.\text{i.m.}}\, g_\alpha(\omega) = g(\omega)$ if

$$\lim_{\alpha\to\alpha_0} \int_{-\infty}^{\infty} |g_\alpha(\omega) - g(\omega)|^2\, d\omega = 0,$$

where α_0 is any fixed constant including ∞. Because of our assumption for f(t) in (3.1), clearly $S(\omega)$ in (3.2) exists and so does the difference $S(\omega+\varepsilon) - S(\omega-\varepsilon)$ in (3.3), which is the ordinary Fourier transform of $f(t)\ \frac{2 \sin \varepsilon t}{t}$. Here $S(\omega)$ defined by (3.2) is termed as the *generalized Fourier transform* of f(t).

At this stage, Wiener uses a Tauberian theorem: If $\varphi(t) \geqslant 0$ for $0 \leqslant t < \infty$, and either of the limits

$$\lim_{T\to\infty} \frac{1}{T} \int_0^T \varphi(t)\, dt \quad \text{or}$$

$$\lim_{\varepsilon\to 0} \frac{2}{\pi\varepsilon} \int_0^\infty \varphi(t) \frac{\sin^2 \varepsilon t}{t^2}\, dt$$

exists, then the other limit exists and assumes the same value. Wiener's proof of this theorem is very complicated. However as a corollary, we have the result that

$$\lim_{\varepsilon\to 0} \frac{1}{4\pi\varepsilon} \int_{-\infty}^{\infty} |S(\omega+\varepsilon) - S(\omega-\varepsilon)|^2\, d\omega$$
$$= \lim_{T\to\infty} \frac{1}{2T} \int_{-T}^{T} |f(t)|^2\, dt.$$

We now define $S_\tau(\omega)$ which bears the same relation to $f(t + \tau)$ as $S(\omega)$ to f(t) in (3.2). Then we have

$$S_\tau(\omega+\varepsilon) - S_\tau(\omega-\varepsilon) - e^{i\tau\omega} \left\{S(\omega+\varepsilon) - S(\omega-\varepsilon)\right\}$$
$$= \frac{1}{\sqrt{2\pi}} \left\{ \underset{A\to\infty}{\text{l.i.m.}} \int_{-A}^{A} f(t + \tau) \frac{2 \sin \varepsilon t}{t} e^{-i\omega t}\, dt \right.$$
$$\left. - \underset{A\to\infty}{\text{l.i.m.}} \int_{-A}^{A} f(t) \frac{2 \sin \varepsilon t}{t} e^{-i\omega(t-\tau)}\, dt \right\}.$$

This can be transformed into

$$\underset{A\to\infty}{\text{l.i.m.}} \frac{1}{\sqrt{2\pi}} \int_{-A}^{A} f(t) \left[\frac{2 \sin \varepsilon(t-\tau)}{t - \tau} - \frac{2 \sin \varepsilon t}{t}\right] e^{-i\omega(t-\tau)}\, dt.$$

Thus, from the Plancherel theorem, it follows that

$$\int_{-\infty}^{\infty} \left| S_\tau(\omega+\varepsilon) - S_\tau(\omega-\varepsilon) - e^{i\omega\tau} \left\{ S(\omega+\varepsilon) - S(\omega-\varepsilon) \right\} \right|^2 d\omega$$

$$= \int_{-\infty}^{\infty} |f(t)|^2 \left[\frac{2 \sin (t-\tau)\varepsilon}{t - \tau} - \frac{2 \sin t\varepsilon}{t} \right]^2 dt.$$

Now it can be shown that

$$\left| \frac{2 \sin (t-\tau)\varepsilon}{t - \tau} - \frac{2 \sin t\varepsilon}{t} \right| \leqslant \frac{16\varepsilon|\tau|}{|t| + |\tau|},$$

and hence by the Tauberian theorem and the fact that if $\frac{1}{2T} \int_{-T}^{T} |f(t)|^2 \, dt$ is bounded in T, then $\int_{-\infty}^{\infty} (|f(t)|^2/1+t^2) \, dt < \infty$, we have

$$\int_{-\infty}^{\infty} \left| S_\tau(\omega+\varepsilon) - S_\tau(\omega-\varepsilon) - e^{i\tau\omega} \left\{ S(\omega+\varepsilon) - S(\omega-\varepsilon) \right\} \right|^2 d\omega = O(\varepsilon^2).$$

Then from this, Wiener derives that

$$\varphi(\tau) = \lim_{\varepsilon\to 0} \frac{1}{4\pi\varepsilon} \int_{-\infty}^{\infty} e^{i\omega\tau} \, |S(\omega+\varepsilon) - S(\omega-\varepsilon)|^2 \, d\omega. \tag{3.4}$$

Putting

$$\varphi_\varepsilon(\tau) = \frac{1}{4\pi\varepsilon} \int_{-\infty}^{\infty} e^{i\omega\tau} \, |S(\omega+\varepsilon) - S(\omega-\varepsilon)|^2 \, d\omega,$$

Wiener shows that the function

$$\sigma_\varepsilon(\omega) := \frac{1}{\sqrt{2\pi}} \left\{ \underset{A\to\infty}{\ell.i.m.} \left[\int_1^A + \int_{-A}^{-1} \right] \frac{\varphi_\varepsilon(t) \, e^{-i\omega t}}{-it} \, dt + \int_{-1}^{1} \varphi_\varepsilon(t) \, \frac{e^{-i\omega t} - 1}{-it} \, dt \right\}$$

is equal to

$$\text{constant} + \int_0^{\omega} \frac{1}{2\varepsilon\sqrt{2\pi}} \, |S(\omega'+\varepsilon) - S(\omega'-\varepsilon)|^2 \, d\omega'.$$

Or, with $\sigma(\omega) = \underset{\varepsilon\to 0}{\ell.i.m.}\, \sigma_\varepsilon(\omega)$, we have

$$\sigma(\omega) - \sigma(-\omega) = \underset{\varepsilon\to 0}{\ell.i.m.} \frac{1}{2\varepsilon\sqrt{2\pi}} \int_{-\omega}^{\omega} |S(\omega'+\varepsilon) - S(\omega'-\varepsilon)|^2 \, d\omega'. \tag{3.5}$$

Finally, Wiener shows that

$$\varphi(\tau) = \frac{1}{\sqrt{2\pi}} \int_{-\infty}^{\infty} e^{i\tau\omega} \, d\sigma(\omega). \tag{3.6}$$

If we let $F(\omega) = \frac{1}{\sqrt{2\pi}} (\sigma(\omega) - \sigma(-\omega))$, then equation (3.6) can be written as

$$\varphi(\tau) = \int_{0}^{\infty} \cos \omega\tau \, dF(\omega). \tag{3.7}$$

In this way we have introduced the spectral function $F(\omega)$ of $f(t)$ and have established the generalized Fourier representation. It should be noted that this concept just follows from the assumption of the existence of the *auto-correlation function* of $f(t)$ defined by

$$\lim_{T\to\infty} \frac{1}{2T} \int_{-T}^{T} f(t + \tau) f^*(t) \, dt,$$

which has been denoted by $\varphi(\tau)$ (cf. (3.1)). This auto-correlation function $\varphi(\tau)$ is found by a summation of contributions of the different frequencies in the spectrum. This leads to consider the function $f(t)$ as a realization of a stochastic process, consisting of superposition of different harmonic waves, the amplitudes of which are random variables of ω.

We first remark that if $f(t)$ is stationary stochastic process of the second order*, with the understanding that the process

---

*) A stochastic process $X(t)$ is said to be of second order if $E\{X^2(t)\} < \infty$ for all t in its domain of definition; the covariance function of $X(t)$ is defined by $E\{[X(\tau) - m(\tau)] [X(t) - m(t)]^*\}$, where $m(t) = E\{X(t)\}$.

is a centered process, i.e. $E\{f(t)\} = 0$, there exists a covariance function $E\{f(t+\tau)f^*(t)\}$ which is independent of t for a stationary process. If the process is *ergodic*, then this covariance function is equal to the auto-correlation function $\varphi(\tau)$, i.e.

$$E\{f(t+\tau)f^*(t)\} = \lim_{T\to\infty} \frac{1}{2T}\int_{-T}^{T} f(t+\tau)f^*(t)\,dt. \tag{3.8}$$

This means that from the relation (3.7), we can consider the process f(t) as the superposition of the stochastic process, $A(\omega)$, which are functions of the frequency ω with independent increments in such a way that $E\{dA(\omega)dA^*(\omega)\} = 0$, if $d\omega \neq d\omega'$. Thus, if we put $dF(\omega) = 2E\{dA(\omega)dA^*(\omega)\}$, we can express f(t) as

$$f(t) = \int_{-\infty}^{\infty} e^{i\omega t}\,dA(\omega). \tag{3.9}$$

In the case of the surface wave, we have a vector $\underset{\sim}{x} = (x,y)$ and the representation takes the form of a superposition of plane waves as a stochastic integral:

$$\eta(x,y,t) = \int_{-\infty}^{\infty}\int e^{i(\underset{\sim}{k}\cdot\underset{\sim}{x} - \omega t)}\,dA(\underset{\sim}{k}), \tag{3.10}$$

where ω and $\underset{\sim}{k} = (k_1,k_2)$ are related by the dispersion relation $\omega = H(\underset{\sim}{k})$. The energy spectrum is given by $dF = E\{dA(\underset{\sim}{k})dA^*(\underset{\sim}{k})\}$. The spectrum depends on both the wave number $k = \sqrt{k_1^2 + k_2^2}$ and the direction ϑ such that $k_1 = k\cos\vartheta$ and $k_2 = k\sin\vartheta$. We denote it by $dF(k,\vartheta)$.

In Chapter 4 we will consider the response of a ship to this incoming wave (3.10). We will see that for a purely sinusoidal incoming wave, the ship motion is also purely sinusoidal in its six degrees of freedom and there exists a transfer matrix $\mu = (\mu_k\mu_j^*)$ with the entries $\mu_k\mu_j^*$ depending on ω (or k) such that $\alpha_k = \mu_k A$, where α_k is defined by (4.19). If $dA(\omega)$ is a stochastic input function, then the α_k will also be stochastic. From the energy spectrum $dF = E\{dAdA^*\}$ of the incoming wave, we can form the covariance matrix of the motion of the ship, $E\{d\alpha_k d\alpha_j^*\}$. The diagonal elements are called the *quadrature spectrum* of the corresponding motions; the non-diagonal elements form the *cross spectrum*. It is easy to see that

$$\begin{aligned} E\{d\alpha_k d\alpha_j^*\} &= E\{\mu_k dA\mu_j^* dA^*\} \\ &= \mu_k\mu_j^*\ E\{dAdA^*\} \qquad (3.11) \\ &= \mu_k\mu_j^*\ dF. \end{aligned}$$

The matrix $\mu_k\mu_j^*$ forms the transfer matrix of the ship and is related to α_k's by the above expression (3.11). Most experiments determine this transfer matrix from measurements of the ship motion with given incoming seas.

3.2 The Shallow Water Theory

In this section we shall discuss the shallow water theory which gives a different kind of approximation from the foregoing linear theory of waves of small amplitude (cf. §1.3). Here it is assumed that the depth of the water is sufficiently small compared with some other characteristic length associated with the horizontal direction.

For convenience, we denote by $(\bar{x},\bar{y},\bar{z})$ and $\bar{t}$ the dimensional space variables and time variable, respectively. The disposition of the coordinate axes is taken in the usual manner, with the $\bar{x},\bar{z}$-plane, the undisturbed water surface and the $\bar{y}$-axis positive upward. The free surface is given by $\bar{y} = \bar{\eta}(\bar{x},\bar{z},\bar{t})$ and a flat bottom is given by $\bar{y} = -h$, $h > 0$ (see Figure 3.1). The velocity components are denoted by $\bar{u}(\bar{x},\bar{y},\bar{z},\bar{t})$, $\bar{v}(\bar{x},\bar{y},\bar{z},\bar{t})$, $\bar{w}(\bar{x},\bar{y},\bar{z},\bar{t})$ and the pressure is denoted by $\bar{p}(\bar{x},\bar{y},\bar{z},\bar{t})$. We recapitulate the basic equations and boundary conditions in terms of the Euler variables $(\bar{x},\bar{y},\bar{z})$ and $\bar{t}$. The equations of motion in (1.1) take the form:

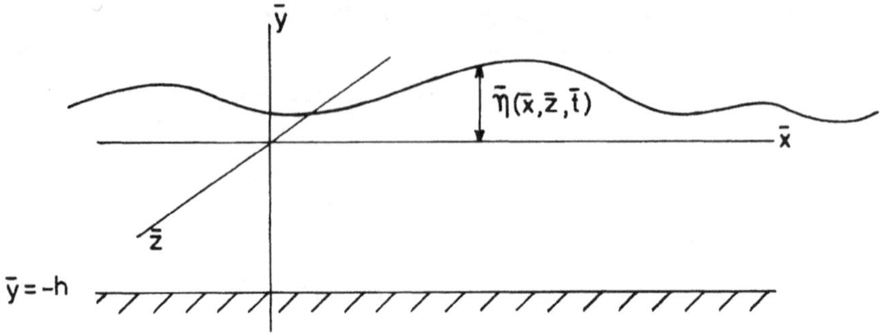

Figure 3.1

$$
\begin{aligned}
\bar{u}_{\bar{t}} + \bar{u}\bar{u}_{\bar{x}} + \bar{v}\bar{u}_{\bar{y}} + \bar{w}\bar{u}_{\bar{z}} &= -\frac{1}{\rho}\,\bar{p}_{\bar{x}}, \\
\bar{v}_{\bar{t}} + \bar{u}\bar{v}_{\bar{x}} + \bar{v}\bar{v}_{\bar{y}} + \bar{w}\bar{v}_{\bar{z}} &= -\frac{1}{\rho}\,\bar{p}_{\bar{y}} - g, \\
\bar{w}_{\bar{t}} + \bar{u}\bar{w}_{\bar{x}} + \bar{v}\bar{w}_{\bar{y}} + \bar{w}\bar{w}_{\bar{z}} &= -\frac{1}{\rho}\,\bar{p}_{\bar{z}}.
\end{aligned}
\tag{3.12}
$$

The equation of continuity is

$$\bar{u}_{\bar{x}} + \bar{v}_{\bar{y}} + \bar{w}_{\bar{z}} = 0 \tag{3.13}$$

and the condition of irrotational flow yields

$$
\begin{aligned}
\bar{u}_{\bar{y}} - \bar{v}_{\bar{x}} &= 0, \\
\bar{v}_{\bar{z}} - \bar{w}_{\bar{y}} &= 0, \\
\bar{w}_{\bar{x}} - \bar{u}_{\bar{z}} &= 0.
\end{aligned}
\tag{3.14}
$$

At the free surface $\bar{y} = \bar{\eta}(\bar{x},\bar{z},\bar{t})$, we have the kinematic condition from (1.11)

$$\bar{\eta}_{\bar{t}} + \bar{u}\bar{\eta}_{\bar{x}} + \bar{w}\bar{\eta}_{\bar{z}} = \bar{v} \tag{3.15}$$

and the condition of constant pressure from (1.13):

$$\Phi_{\bar{t}} + \frac{1}{2}\,(\bar{u}^2 + \bar{v}^2 + \bar{w}^2) + g\bar{\eta} = \text{constant} \tag{3.16}$$

where Φ is the velocity potential which exists because of condition (3.14). By assuming the bottom flat, we have the boundary condition

$$\bar{v} = 0, \quad \text{at } \bar{y} = -h. \tag{3.17}$$

We now introduce dimensionless space variables:

$$x = \frac{\bar{x}}{L}, \qquad z = \frac{\bar{z}}{L}, \qquad y = \frac{\bar{y}}{\bar{h}} = \frac{\bar{y}}{\varepsilon L},$$

where $\varepsilon = \bar{h}/L$ and L is a reference length in $\bar{x}$ and $\bar{z}$ direction. Then by introducing a reference velocity U (which will be chosen later) and putting

$$u = \frac{\bar{u}}{U}, \qquad \varepsilon v = \frac{\bar{v}}{U}, \qquad w = \frac{\bar{w}}{U},$$

the equation of continuity (3.13) reads:

$$u_x + v_y + w_z = 0, \tag{3.18}$$

and the vorticity free conditions (3.14) give

$$\begin{aligned} u_y - \varepsilon^2 v_x &= 0, \\ w_y - \varepsilon^2 v_z &= 0, \\ w_x - \quad u_z &= 0. \end{aligned} \tag{3.19}$$

Taking for the time scale $t = \bar{t}U/L$, we have the equations of motion in the form:

$$\begin{aligned} u_t + uu_x + vu_y + wu_z &= -\frac{1}{\rho U^2}\bar{p}_x, \\ \varepsilon^2(v_t + uv_x + vv_y + wv_z) &= -\frac{1}{\rho U^2}\bar{p}_y - \frac{\varepsilon g L}{U^2}, \\ w_t + uw_x + vw_y + ww_z &= -\frac{1}{\rho U^2}\bar{p}_z. \end{aligned}$$

We now choose U such that $\varepsilon g L = U^2$. Then the above equations become, with $\pi = \bar{p}/\rho U^2$

$$u_t + uu_x + vu_y + wu_z = -\pi_x,$$

$$\varepsilon^2(v_t + uv_x + vv_y + wv_z) = -\pi_y - 1, \tag{3.20}$$

$$w_t + uw_x + vw_y + ww_z = -\pi_z.$$

Similarly introducing the dimensionless variables $\varphi = \Phi/UL$, $\eta = \bar{\eta}/h$ in the equations (3.15) to (3.17) yields

$$\left.\begin{aligned} &\eta_t + u\eta_x + w\eta_z = v && (3.21)\\ &\varphi_t + \tfrac{1}{2}(u^2 + \varepsilon^2 v^2 + w^2) + \eta = \text{constant} && (3.22)\end{aligned}\right\} \text{ at } y = \eta,$$

$$v = 0, \text{ at } y = -1. \tag{3.23}$$

Apparently the relevant parameter is ε^2, hence the object now is to consider solutions of the problem (3.18)-(3.20) together with (3.21)-(3.23), depending on this small parameter and then develop in powers of ε^2.

First we assume power series developments for u, v, w and π:

$$\begin{aligned} u &= u_0 + \varepsilon^2 u_1 + \varepsilon^4 u_2 + \ldots\ldots \\ v &= v_0 + \varepsilon^2 v_1 + \varepsilon^4 v_2 + \ldots\ldots \\ w &= w_0 + \varepsilon^2 w_1 + \varepsilon^4 w_2 + \ldots\ldots \\ \pi &= \pi_0 + \varepsilon^2 \pi_1 + \varepsilon^4 \pi_2 + \ldots\ldots \end{aligned} \tag{3.24}$$

and substitute them into equations (3.18) to (3.20). Equating to zero the coefficients of like powers of ε^2 gives the successive system of equations. The terms of zero order yield the equations:

$$u_{0y} + v_{0y} + w_{0z} = 0 \tag{3.25}$$

$$u_{0y} = 0; \quad w_{0y} = 0; \quad w_{0x} - u_{0z} = 0. \tag{3.26}$$

$$u_{0t} + u_0 u_{0x} + v_0 u_{0y} + w_0 u_{0z} = -\pi_{0x},$$

$$0 = -\pi_{0y} - 1, \qquad (3.27)$$

$$w_{0t} + u_0 w_{0x} + v_0 w_{0y} + w_0 w_{0z} = -\pi_{0z}.$$

Putting $\eta = \eta_0 + \varepsilon^2\eta_1 + \varepsilon^4\eta_2 + \ldots\ldots$, the free surface conditions are

$$v_0 = \eta_{0t} + u_0\eta_{0x} + w_0\eta_{0z} \qquad (3.28)$$

together with

$$\pi_0(\eta) = 0. \qquad (3.29)$$

At the bottom, we have the boundary condition

$$v_0 = 0, \quad \text{at } y = -1, \qquad (3.30)$$

and from the continuity equation (3.25), we then find

$$v_0(\eta) = -\int_{-1}^{\eta} (u_{0x} + w_{0z})\,dy,$$

but from (3.26), since u_0 and w_0 are independent of y, it follows that

$$v_0(\eta) = -(u_{0x} + w_{0z})(\eta + 1)$$

$$= -(u_{0x} + w_{0z})(\eta_0 + 1) + O(\varepsilon^2). \qquad (3.31)$$

Also from the second equation of (3.27), $\pi_{0y} = -1$ and the fact that $\pi_0(\eta) = 0$ from (3.29), we see that to the same order of approximation

$$\pi_0 = \eta_0 - y, \qquad (3.32)$$

which corresponds to a hydrostatic pressure distribution. Collecting results (3.31) and (3.32), we obtain from (3.27) the equations:

$$u_{0t} + u_0 u_{0x} + w_0 u_{0z} + \eta_{0x} = 0,$$
$$w_{0t} + u_0 w_{0x} + w_0 w_{0z} + \eta_{0z} = 0, \tag{3.33}$$

together with

$$\eta_{0t} + u_0 \eta_{0x} + w_0 \eta_{0z} + (\eta_0 + 1)(u_{0x} + w_{0z}) = 0,$$

from (3.28). Observe that the terms $v_0 u_{0y}$ and $v_0 w_{0y}$ in (3.27) vanish because of equation (3.26). Hence we arrive at, in the first approximation, the following set of quasi-linear equations for u, w and the total thickness of the water layer η:

$$u_t + uu_x + wu_z + \eta_x = 0,$$
$$w_t + uw_x + ww_z + \eta_z = 0, \tag{3.34}$$
$$\eta_t + u\eta_x + w\eta_z + (\eta+1)u_x + (\eta+1)w_z = 0.$$

These are the dimensionless forms of the basic equations of shallow water theory.

In the following, we shall consider two special cases of (3.34) which are of importance: the steady case from (3.26) and (3.34):

$$w_x - u_z = 0,$$
$$uu_x + wu_z + \eta_x = 0,$$
$$uw_x + ww_z + \eta_z = 0, \tag{3.35}$$
$$u\eta_x + w\eta_z + (\eta+1)u_x + (\eta+1)w_z = 0,$$

and the one-dimensional non-steady case from (3.34):

$$u_t + uu_x + \eta_x = 0,$$
$$\eta_t + u\eta_x + (\eta+1)u_x = 0. \tag{3.36}$$

In the first case, from the first three equations of (3.35), we find that

$$\frac{\partial}{\partial x}\left\{\frac{1}{2}(u^2 + w^2) + \eta\right\} = 0,$$

$$\frac{\partial}{\partial z}\left\{\frac{1}{2}(u^2 + w^2) + \eta\right\} = 0,$$

or

$$\eta = \frac{1}{2}\left\{(c_0 - 2) - (u^2 + w^2)\right\}, \qquad (3.37)$$

$$c_0 = \text{constant}.$$

Substitution (3.37) into the last equation of (3.35) gives a second order quasi-linear equation for the potential function φ:

$$\frac{\partial}{\partial x}\left\{\left[c_0 - (u^2 + w^2)\right]\varphi_x\right\} + \frac{\partial}{\partial z}\left\{\left[c_0 - (u^2 + w^2)\right]\varphi_z\right\} = 0,$$

or

$$\left[c_0 - (3u^2 + w^2)\right]\varphi_{xx} - 4uw\varphi_{xz} + \left[c_0 - (u^2 + 3w^2)\right]\varphi_{zz} = 0. \qquad (3.38)$$

The steady flow in this case is completely analogous to the steady two-dimensional gas flow.

In the second case, the unsteady one-dimensional case, the equations (3.36) can be solved by the method of characteristics. We define here the characteristic directions α, β as those for which a linear combination of the equations yields differentiation of the functions η and u in the same direction. A linear combination of the equations (3.36) is

$$\lambda u_t + (\lambda u + \mu(\eta+1))\, u_x + \mu\eta_t + (\lambda + \mu u)\eta_x = 0, \qquad (3.39)$$

and the differentiations of u and η are in the same direction if

$$\frac{\lambda u + \mu(\eta+1)}{\lambda} = \frac{\lambda + \mu u}{\mu},$$

which gives $\lambda = \mu\sqrt{\eta + 1}$ or $\lambda = -\mu\sqrt{\eta + 1}$.
Hence we have finally from (3.39), putting $c = \sqrt{\eta + 1}$:

$$\begin{aligned} cu_t + c(u + c)u_x + \eta_t + (u + c)\eta_x &= 0, \\ -cu_t - c(u - c)u_x + \eta_t + (u - c)\eta_x &= 0. \end{aligned} \tag{3.40}$$

The characteristic directions are

$$\begin{aligned} \frac{\partial}{\partial\alpha} &= \frac{\partial}{\partial t} + (u + c)\frac{\partial}{\partial x}, \\ \frac{\partial}{\partial\beta} &= \frac{\partial}{\partial t} + (u - c)\frac{\partial}{\partial x}, \end{aligned} \tag{3.41}$$

and the equations (3.40) take the form:

$$c\frac{\partial u}{\partial\alpha} + \frac{\partial\eta}{\partial\alpha} = 0 \text{ and}$$

$$-c\frac{\partial u}{\partial\beta} + \frac{\partial\eta}{\partial\beta} = 0$$

or (3.42)

$$\frac{\partial u}{\partial\alpha} + 2\frac{\partial(\eta+1)^{\frac{1}{2}}}{\partial\alpha} \text{ and}$$

$$-\frac{\partial u}{\partial\beta} + 2\frac{\partial(\eta+1)^{\frac{1}{2}}}{\partial\beta} = 0,$$

by making use of $c = \sqrt{\eta + 1}$. This gives the Riemann invariants:

$$\begin{aligned} u + 2(\eta+1)^{\frac{1}{2}} &= f(\beta), \\ -u + 2(\eta+1)^{\frac{1}{2}} &= g(\alpha). \end{aligned} \tag{3.43}$$

An application of the above theory to the breaking of a dam can be found in Stoker's book.

3.3 Non-linear Dispersive Waves

As we have seen in the previous section, we can solve the shallow water equations, in the first approximation, by the method of characteristics. If these characteristics intersect, a shock will develop, which is not included in the equation of the first approximation. For this reason, it has an advantage to consider the second approximation to the equations.

We shall derive the second order approximation for the one-dimensional flow by a method which differs slightly from the method used in the previous section, but which is somewhat shorter.

Again, we denote the dimensional coordinates system by $(\bar{x},\bar{y})$, but for convenience, we place the $\bar{x}$-axis on the flat bottom as shown in Figure 3.2. The boundary value problem for Φ consists of equations:

$$\begin{aligned} &\Phi_{\bar{x}\bar{x}} + \Phi_{\bar{y}\bar{y}} = 0, \quad \text{for} \begin{cases} -\infty < x < \infty \\ 0 < y < \bar{\eta}(\bar{x},\bar{t}) \end{cases} \\ &\Phi_{\bar{y}} = 0, \quad \text{at } \bar{y} = 0, \end{aligned} \tag{3.44}$$

and the following conditions on the free surface $\bar{y} = \bar{\eta}(\bar{x},\bar{t})$. On the free surface, the kinematic condition (3.15)

$$\bar{\eta}_{\bar{t}} + \bar{u}\bar{\eta}_{\bar{x}} = \bar{v}, \tag{3.45}$$

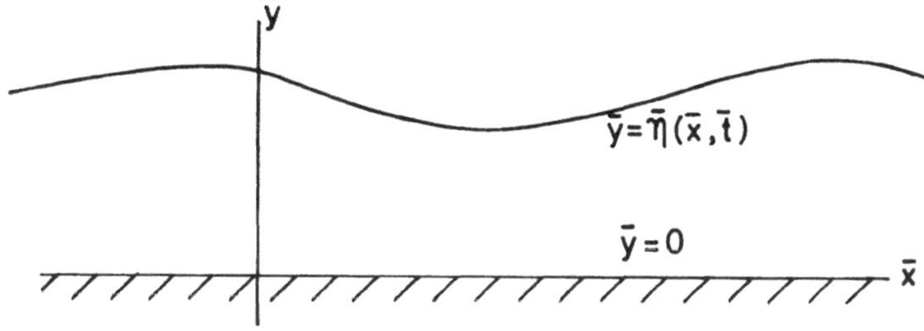

Figure 3.2

holds together with the dynamic condition of constant pressure $\bar{p}$, which takes the form $\partial\bar{p}/\partial s = 0$. Here $\partial/\partial s$ denotes differentiation in the direction of the surface face. Since the tangent to the free surface has direction cosines $(1, \bar{\eta}_{\bar{x}})$, this dynamic condition on the free surface becomes

$$\frac{\partial \bar{p}}{\partial \bar{x}} + \frac{\partial \bar{p}}{\partial \bar{y}} \bar{\eta}_{\bar{x}} = 0, \tag{3.46}$$

where $\frac{\partial \bar{p}}{\partial \bar{x}}$ and $\frac{\partial \bar{p}}{\partial \bar{y}}$ follow from the equations of motion, that is:

$$\begin{aligned} \bar{u}_{\bar{t}} + \bar{u}\bar{u}_{\bar{x}} + \bar{v}\bar{u}_{\bar{y}} &= -\frac{1}{\rho}\bar{p}_{\bar{x}}, \\ \bar{v}_{\bar{t}} + \bar{u}\bar{v}_{\bar{x}} + \bar{v}\bar{v}_{\bar{y}} &= -\frac{1}{\rho}\bar{p}_{\bar{y}} - g. \end{aligned} \tag{3.47}$$

Substitution of the above equations into (3.46) gives

$$\bar{u}_{\bar{t}} + \bar{u}\bar{u}_{\bar{x}} + \bar{v}\bar{u}_{\bar{y}} + \bar{\eta}_{\bar{x}}(\bar{v}_{\bar{t}} + \bar{u}\bar{v}_{\bar{x}} + \bar{v}\bar{v}_{\bar{y}} + g) = 0, \tag{3.48}$$

(cf. (1.18)).

In order to obtain the shallow water equations (3.33), we consider the complex velocity $\bar{w}$ defined by $\bar{w} = \bar{u} - i\bar{v}$ which is an analytic function of the complex variable $\bar{x} + i\bar{y}$ (or more precisely $\bar{x} + i\bar{y}$ and $\bar{t}$) in the interior of the region occupied by the fluid. We will assume this regularity continues to remain valid up to and including the boundary of the region. Then expanding $\bar{w}$ into a Taylor expansion about $\bar{x}$ for fixed $\bar{t}$, we find

$$\bar{u} - i\bar{v} = \bar{w}(\bar{x}) + i\bar{y}\,\bar{w}'(x) - \frac{\bar{y}^2}{2}\bar{w}''(\bar{x}) - \frac{i}{6}\bar{y}^3\,\bar{w}'''(\bar{x}) + \ldots\ldots \tag{3.49}$$

(Here for simplicity, we omit the argument $\bar{t}$ and henceforth in the sequel.) Since at the flat bottom $\bar{y} = 0$, we have

$$\bar{v} = \bar{v}_{\bar{x}} = \bar{v}_{\bar{x}\bar{x}} = \bar{v}_{\bar{x}\bar{x}\bar{x}} = \ldots\ldots = 0, \tag{3.50}$$

it follows that by differentiating both sides of (3.49) with respect to $\bar{x}$, we obtain for $\bar{y} = 0$:

$$\begin{aligned} \bar{w}(\bar{x}) &= \bar{u}(\bar{x},0), \\ \bar{w}'(\bar{x}) &= \bar{u}_{\bar{x}}(\bar{x},0), \\ \bar{w}''(\bar{x}) &= \bar{u}_{\bar{x}\bar{x}}(\bar{x},0), \ \ldots\ldots \end{aligned} \tag{3.51}$$

This gives the expansions:

$$\begin{aligned} \bar{u}(\bar{x},\bar{y}) &= \bar{u}(\bar{x},0) - \frac{\bar{y}^2}{2}\,\bar{u}_{\bar{x}\bar{x}}(\bar{x},0) + \frac{\bar{y}^4}{4!}\,\bar{u}_{\bar{x}\bar{x}\bar{x}\bar{x}}(\bar{x},0) \\ &\quad + \ldots\ldots \\ \bar{v}(\bar{x},\bar{y}) &= -\bar{y}\bar{u}_{\bar{x}}(\bar{x},0) + \frac{\bar{y}^3}{3!}\,\bar{u}_{\bar{x}\bar{x}\bar{x}}(\bar{x},0) \\ &\quad - \frac{\bar{y}^5}{5!}\,\bar{u}_{\bar{x}\bar{x}\bar{x}\bar{x}\bar{x}}(\bar{x},0) + \ldots\ldots \end{aligned} \tag{3.52}$$

We now introduce dimensionless coordinates by putting $x = \bar{x}/L$, $y = \bar{y}/h$, where h is the characteristic length of the water depth and L is the 'wave length' in the $\bar{x}$-direction. Moreover, we put $\bar{u} = Uu$ with U being a reference velocity. The time $\bar{t}$ is made dimensionless by the introduction of a time scale T such that $t = \bar{t}/T$. Putting $\delta = h/L$, we see that $\bar{v}$ is of order δ with respect to $\bar{u}$. We put $\bar{v} = \delta Uv$. Furthermore, we refer to $u(x,0,t)$ as

$$u(x,0,t) = f(x,t). \tag{3.53}$$

Then from (3.52), we have the expansions:

$$\begin{aligned} u(x,y,t) &= f - \frac{\delta^2}{2}\,y^2 f_{xx} + \frac{\delta^4}{4!}\,y^4 f_{xxxx} + O(\delta^6) \\ v(x,y,t) &= -yf_x + \frac{\delta^2}{3!}\,y^3 f_{xxx} - \frac{\delta^4}{5!}\,y^5 f_{xxxx} \\ &\quad + O(\delta^6). \end{aligned} \tag{3.54}$$

On the surface $y = \eta(x,t)$, $\eta = \bar{\eta}/h$, the kinematic condition

(3.45) takes the form:

$$\frac{h}{T}\eta_t + \frac{Uh}{L}u\eta_x = U\delta v, \tag{3.55}$$

or by setting $\delta^s = L/UT$ for some non-negative power s, it becomes that

$$\delta^s\eta_t + u\eta_x = v. \tag{3.56}$$

The constant pressure condition (3.48) is transformed into

$$\delta^s u_t + uu_x + vu_y + \delta^2\eta_x\left(\delta^s v_t + uv_x + vv_y\right) + \eta_x \frac{gL\delta}{U^2} = 0. \tag{3.57}$$

We choose U such that $U^2 = gL\delta$ and substitute the equations (3.54) for u and v into (3.56) and (3.57). Then we obtain a set of equations for f and η:

$$\delta^s\eta_t + (f\eta_x + \eta f_x) - \frac{1}{2}\delta^2\eta^2\left(\eta_x f_{xx} + \frac{1}{3}\eta f_{xxx}\right) + \frac{\delta^4}{4!}\eta^4\left(\eta_x f_{xxxx} + \frac{1}{5}\eta f_{xxxxx}\right) + O(\delta^6) = 0. \tag{3.58}$$

$$\begin{aligned}\delta^s\Big[f_t &- \delta^2\left(\frac{\eta^2}{2}f_{xxt} + \eta\eta_x f_{xt}\right) + \\ &+ \frac{\delta^4}{6}\eta^3\left(\eta_x f_{xxxt} + \frac{1}{4}\eta f_{xxxxt}\right)\Big] + \left(ff_x + \eta_x\right) \\ &+ \delta^2\eta\left(\eta_x f_x^2 - \eta_x ff_{xx} + \frac{1}{2}\eta f_x f_{xx} - \frac{1}{2}\eta ff_{xxx}\right) \\ &+ \delta^4\eta^3\left[\frac{\eta}{4}\left(\frac{1}{3}f_{xx}f_{xxx} - \frac{1}{2}f_x f_{xxxx} + \frac{1}{6}ff_{xxxxx}\right)\right. \\ &\left.+ \eta_x\left(\frac{1}{2}f_{xx}^2 - \frac{2}{3}f_x f_{xxx} + \frac{1}{6}ff_{xxxx}\right)\right] + O(\delta^6) = 0.\end{aligned} \tag{3.59}$$

We construct asymptotic solutions of this set of equations by substituting for f and η the power series in δ^2:

$$f = u_0 + \delta^2 u_1 + \delta^4 u_2 + \ldots\ldots$$
$$\eta = \eta_0 + \delta^2 \eta_1 + \delta^4 \eta_2 + \ldots\ldots \tag{3.60}$$

There are two cases to be considered:

a. CASE 1: $s = 0$

In this case the time scale T is adapted to the reference speed U by the relation $U = L/T$. Substitution of the series (3.60) into (3.58) and (3.59) gives, after equating to zero the coefficients of like powers of δ^2, the first approximation:

$$\begin{aligned} \eta_{0t} + u_0\eta_{0x} + \eta_0 u_{0x} &= 0, \\ u_{0t} + u_0 u_{0x} + \eta_{0x} &= 0, \end{aligned} \tag{3.61}$$

which are the shallow water equations for one-dimensional flow (3.33).
The second approximation leads to the following:

$$\begin{aligned} &\eta_{1t} + \eta_0 u_{1x} + \eta_1 u_{0x} + \eta_{0x} u_1 + \eta_{1x} u_0 \\ &\quad - \frac{\eta_0^2}{6}\left(\eta_0 u_{0xxx} + 3\eta_{0x} u_{0xx}\right) = 0, \\ &u_{1t} + \eta_{1x} + u_0 u_{1x} + u_{0x} u_1 - \frac{1}{2}\eta_0^2 u_{0xxt} \\ &\quad - \frac{\eta_0^2}{2}\left(u_0 u_{0xxx} - u_{0x} u_{0xx}\right) - \eta_0 \eta_{0x} u_{0xt} \\ &\quad - \eta_0 \eta_{0x} u_0 u_{0xx} + \eta_0 \eta_{0x} u_{0x}^2 = 0. \end{aligned} \tag{3.62}$$

We consider a special case of the first approximation due to a flow with constant velocity which we put equal to one at $y = 1$. In this case since (3.62) gives only the trivial solutions, we must consider an approximation somewhat different than the ones used in (3.62). We reexamine (3.58) and (3.59) by considering the expansions:

$$f = 1 + \delta^2 u_1 + \delta^4 u_2 + \ldots\ldots$$
$$\eta = 1 + \delta^2 \eta_1 + \delta^4 \eta_2 + \ldots\ldots \tag{3.63}$$

We keep all terms up to order δ^4 except those terms involving u_2 and η_2, and obtain

$$\eta_{1t} + \eta_{1x} + u_{1x} + \delta^2(u_1\eta_{1x} + \eta_1 u_{1x}) - \frac{1}{6}\delta^2 u_{1xxx} = 0,$$
$$u_{1t} + u_{1x} + \eta_{1x} + \delta^2 u_1 u_{1x} - \frac{1}{2}\delta^2(u_{1xxt} + u_{1xxx}) = 0. \tag{3.64}$$

By setting $u = 1 + \delta^2 u_1$ and $\eta = 1 + \delta^2\eta_1$ (or equivalently, $u_1 = \frac{1}{\delta^2}(u - 1)$ and $\eta_1 = \frac{1}{\delta^2}(\eta - 1)$), we obtain from (3.64) a variant of the equations of Boussinesq

$$\eta_t + \eta u_x + u\eta_x - \frac{1}{6}\delta^2 u_{xxx} = 0,$$
$$u_t + uu_x + \eta_x - \frac{1}{2}\delta^2(u_{xxt} + u_{xxx}) = 0. \tag{3.65}$$

The original form of the equations of Boussinesq can be obtained by the following transformation which is allowable in the order of approximation; we first replace u by

$$u = 1 + \delta^2\left(u_1 - \frac{1}{6}\delta^2 u_{1xx}\right), \tag{3.66}$$

and introducing an x'-coordinate system which moves with the main flow $U = 1(x' = x - Ut)$, we then put $u'(x',t) = u(x' + Ut,t) - U$, $\eta'(x',t) = \eta(x' + Ut,t)$. This leads successively to the equations, after dropping the prime:

$$\eta_t + \eta u_x + u\eta_x = 0,$$
$$u_t + uu_x + \eta_x - \frac{1}{3}\delta^2 u_{xxt} = 0. \tag{3.67}$$

Equations (3.67) are known as the *Boussinesq equations* for the one-dimensional shallow water waves.

b. CASE 2: $s = 2$

The first approximation now does not contain t, and we simply have

$$\begin{aligned} u_0\eta_{0x} + \eta_0 u_{0x} &= 0, \\ u_0 u_{0x} + \eta_{0x} &= 0. \end{aligned} \tag{3.68}$$

This is solved by $u_0 = 1$ and $\eta_0 = 1$, if we assume again the series expansions (3.63). Substitution into (3.58) and (3.59) gives the second approximation, the coefficient of δ^2:

$$\begin{aligned} \eta_{1x} + u_{1x} &= 0, \\ u_{1x} + \eta_{1x} &= 0, \end{aligned} \tag{3.69}$$

which shows that $u_1 = -\eta_1 + X(t)$ for some function $X(t)$. If we assume that flow is undisturbed at infinity, then this implies

$$u_1 = -\eta_1 + \alpha \tag{3.70}$$

for some constant α.

Consider now the next approximation, the coefficient of δ^4:

$$\begin{aligned} \eta_{1t} + u_1\eta_{1x} + \eta_1 u_{1x} - \tfrac{1}{6} u_{1xxx} + u_{2x} + \eta_{2x} &= 0, \\ u_{1t} + u_1 u_{1x} - \tfrac{1}{2} u_{1xxx} + \eta_{2x} + u_{2x} &= 0. \end{aligned} \tag{3.71}$$

Subtracting the two equations in order to eliminate u_2 and η_2 gives

$$\eta_{1t} - u_{1t} + u_1(\eta_{1x} - u_{1x}) + \eta_1 u_{1x} + \tfrac{1}{3} u_{1xxx} = 0. \tag{3.72}$$

By using the relation (3.70), we then reduce to an equation of η_1, the *Korteweg-De Vries* equation:

$$\eta_{1t} + \left(\alpha - \tfrac{3}{2}\eta_1\right)\eta_{1x} - \tfrac{1}{6}\eta_{1xxx} = 0. \tag{3.73}$$

In order to find a wave solution of the Korteweg-De Vries equation (3.73), we first change the coordinates in such a way so that the equation takes the form:

$$\eta_t + (1 + \varepsilon\eta)\eta_x + \mu\eta_{xxx} = 0, \qquad (3.74)$$

where ε and μ are small parameters depending upon α. Introducing a large parameter K, we substitute

$$\eta(x,t) = U\left[KS(x,t),\ x,t\right] + \frac{1}{K} V\left[KS(x,t),\ x,t\right] + O\left(\frac{1}{K^2}\right) \qquad (3.75)$$

into equation (3.74) and with the abbreviation $p = KS(x,t)$, we obtain the approximation up to terms of order K:

$$(k - \omega)U_p + \varepsilon kUU_p + (\mu K^2)k^3U_{ppp} + \frac{1}{K} O(\mu K^2) = 0, \qquad (3.76)$$

where we have introduced

$$k = S_x(x,t), \qquad \omega = -S_t(x,t)$$

as the wave number and frequency, respectively.

Now identify K^2 so that $\mu K^2 = 1$ and notice that S is slowly varying with x and t. With this choice we neglect the term $\frac{1}{K} O(\mu K^2)$ in (3.76), since it is of lower order. Then the equation (3.76) is just an ordinary third order differential equation for U as a function of p with coefficients which vary with x and t. This gives after integration with respect to p,

$$(k - \omega)U + \frac{1}{2} \varepsilon kU^2 + k^3U_{pp} = \frac{1}{2} \alpha. \qquad (3.77)$$

Integrating once more after multiplication with U_p yields

$$k^3U_p^2 = \alpha U + \beta + (\omega - k)U^2 - \frac{1}{3} \varepsilon kU^3. \qquad (3.78)$$

Here the constants of integration α and β are functions of x and t.

The equation (3.78) is similar to the equation for the pendulum and it can be seen that U is a periodic function of p oscillating between two zeroes of the right hand side of equation (3.78). In general, this right hand side has three zeroes but we have to select those two zeroes for which the right hand side of (3.78) is positive between these two zeroes.

We now perform a linear substitution:

$$U(p,x,t) = m(x,t) + A(x,t)\,\Theta(p), \tag{3.79}$$

which makes the zeroes of the new function Θ between -1 and +1. Using the fact that $\Theta_p = 0$ for both $\Theta = -1$ and $\Theta = +1$, from (3.78) we can replace α and β by the constants (functions of x and t) m and A, and obtain the equation

$$k^3\Theta_p^2 = (1 - \Theta^2)\ (\tfrac{1}{3}\ \varepsilon Ak\Theta + \varepsilon km + (k - \omega)). \tag{3.80}$$

The dependence of Θ on p can be given in implicit form as:

$$p + \gamma = \sqrt{k^3}\int_{-1}^{\Theta} \frac{d\Theta}{\sqrt{(1 - \Theta^2)(\frac{1}{3}\ \varepsilon Ak\Theta + \varepsilon km + (k - \omega))}}, \tag{3.81}$$

with $\gamma(x,t)$ as a shifting constant. The period of Θ can be normalized to unity by changing the scale of p so that

$$\frac{1}{2} = \sqrt{k^3}\int_{-1}^{1} \frac{d\Theta}{\sqrt{(1 - \Theta^2)(\frac{1}{3}\ \varepsilon Ak\Theta + \varepsilon km + (k - \omega))}}. \tag{3.82}$$

This gives a dispersion relation between k, ω, A and m. Observe that for $\varepsilon = 0$, we simply get the relation

$$\frac{1}{2} = \sqrt{\frac{k^3}{k - \omega}}\int_{-1}^{1} \frac{d\Theta}{\sqrt{1 - \Theta^2}} = \pi\sqrt{\frac{k^3}{k - \omega}}, \tag{3.83}$$

which corresponds to the dispersion relation for the linear equation for cnoidal waves, i.e., the equation (3.74) with

$\varepsilon = 0$.

A second relation between k and ω is the conservation law for wave crests:

$$\frac{\partial \omega}{\partial x} + \frac{\partial k}{\partial t} = 0. \tag{3.84}$$

In order to find two more relations for k, ω, m, and A we return to the original equation (3.74) and substitute into the equation the expression (3.79) for η(x,t), i.e.,

$$\eta(x,t) = m(x,t) + A(x,t)\,\Theta(p).$$

Since

$$\begin{aligned}
\eta_t &= m_t + A_t\Theta - KA\Theta_p\omega, \\
\eta_x &= m_x + A_x\Theta + KA\Theta_p k, \\
\eta_{xx} &= m_{xx} + A_{xx}\Theta + 2KA_x\Theta_p k + KA\Theta_p k_x + K^2A\Theta_{pp}k^2, \\
\eta_{xxx} &= m_{xxx} + A_{xxx}\Theta + 3KA_{xx}\Theta_p k + 3KA_x\Theta_p k_x \\
&\quad + KA\Theta_p k_{xx} + 3K^2A_x\Theta_{pp}k^2 + 3K^2A\Theta_{pp}kk_x \\
&\quad + K^3A\Theta_{ppp}k^3,
\end{aligned}$$

this substitution gives the coefficient of K, (recall $\mu K^2 = 1$),

$$k^3A\Theta_{ppp} + A\Theta_p(k - \omega) + \varepsilon(m + A\Theta)\,A\Theta_p k, \tag{3.85}$$

which vanishes as we will now see. Recall that Θ is determined by (3.80):

$$k^3\Theta_p^2 = (1 - \Theta^2)\left(\tfrac{1}{3}\,\varepsilon Ak\Theta + \varepsilon km + (k - \omega)\right).$$

By differentiating this equation with respect to p, we then obtain, with some reduction, the equation

$$k^3\Theta_{pp} = \frac{1}{6}\,\varepsilon Ak - \Theta(\varepsilon km + (k - \omega)) - \frac{1}{2}\,\varepsilon Ak\Theta^2. \qquad (3.86)$$

Differentiating it once more, we arrive at the equation:

$$k^3\Theta_{ppp} = (\varepsilon km + (k - \omega))\Theta_p - Ak\varepsilon\Theta\Theta_p, \qquad (3.87)$$

which shows that, as was expected, the coefficient of K^1 (cf. 3.85) vanishes.

We expect the solution to be valid for a large number of periods (order K). This requires that the coefficient of K^0 shall not grow after many periods. The coefficient of K^0

$$m_t + A_t\Theta + m_x + A_x\Theta + (m + A\Theta)(m_x + A_x\Theta) + 3\left(A_xk^2 + Akk_x\right)\Theta_{pp}, \qquad (3.88)$$

will be integrated over one period and the integral is put to zero. This gives

$$m_t + A_t\int_0^1 \Theta dp + m_x + A_x\int_0^1 \Theta dp + \varepsilon m\left(m_x + A_x\int_0^1 \Theta dp\right) + \varepsilon Am_x\int_0^1 \Theta dp + \varepsilon AA_x\int_0^1 \Theta^2 dp + 3\left(A_xk^2 + Akk_x\right) \times \left(\Theta_p(1) - \Theta_p(0)\right) = 0. \qquad (3.89)$$

Note that the last term equals zero because of the periodicity of Θ_p. By setting $\gamma_n = \int_0^1 \Theta^n dp$, we see that equation (3.88) can be written in the form:

$$(m + A\gamma_1)_t + (m + A\gamma_1)_x + \varepsilon(m + A\gamma_1)_x(m + A\gamma_1) + \varepsilon AA_x\left(\gamma_2 - \gamma_1^2\right) = 0. \qquad (3.90)$$

Apparently the term $m + A\gamma_1$ is the mean height of the wave over one period in p, since the period of p is one. Thus, if we let $h = m + A\gamma_1$ be the mean wave height, we obtain the relation:

$$\frac{\partial h}{\partial t} + \frac{\partial}{\partial x}\left(h + \frac{1}{2}\varepsilon h^2\right) + \frac{1}{2}\varepsilon\left(\gamma_2 - \gamma_1^2\right)\frac{\partial A^2}{\partial x} = 0. \tag{3.91}$$

The remaining relation can be found by a similar procedure. Multiplying the equation (3.74) by η and substituting (3.79) for η into the equation, we find the coefficient of K,

$$A(m + A\Theta)(k - \omega)\Theta_p + \varepsilon(m + A\Theta)^2 Ak\Theta_p + A\Theta_{ppp}(m + A\Theta)k^3, \tag{3.92}$$

which is evidently zero from (3.87). The coefficient of K^0 is

$$(m + A\Theta)(m_t + A_t\Theta + m_x + A_x\Theta) + \varepsilon(m + A\Theta)^2(m_x + A_x\Theta)$$
$$+ 3(m + A\Theta)\left(A_x k^2 + Akk_x\right)\Theta_{pp}.$$

Integrating this over one period gives the equation

$$\begin{aligned} m(m_t + m_x) &+ m(A_t + A_x)\gamma_1 + (m_t + m_x)A\gamma_1 + A(A_x + A_t)\gamma_2 \\ &+ \varepsilon m^2 m_x + \varepsilon\left(2mm_x A\gamma_1 + m^2 A_x\gamma_1\right) + \varepsilon A^2 A_x\gamma_3 \\ &+ 2\varepsilon mAA_x\gamma_2 + \varepsilon A^2 m_x\gamma_2 + 3\left(mA_x k^2 + mAkk_x\right) \\ &\times\left(\Theta_p(1) - \Theta_p(0)\right) + 3A\left(A_x k^2 + Akk_x\right)\int_0^1 \Theta\Theta_{pp}dp = 0. \end{aligned} \tag{3.93}$$

Note that from (3.86) we have

$$k^3\int_0^1 \Theta\Theta_{pp}dp = \frac{1}{6}\varepsilon Ak\gamma_1 - (\varepsilon km + (k - \omega))\gamma_2 - \frac{1}{2}\varepsilon Ak\gamma_3.$$

Hence, with $h = m + A\gamma_1$, the mean wave height, we obtain from (3.93) the relation:

$$
\begin{aligned}
& h(h_t + h_x) + A(A_t + A_x)\left(\gamma_2 - \gamma_1^2\right) + \varepsilon h^2 h_x \\
& + 2\varepsilon m A A_x\left(\gamma_2 - \gamma_1^2\right) + \varepsilon A^2 m_x\left(\gamma_2 - \gamma_1^2\right) \\
& + \varepsilon A^2 A_x\left(\gamma_3 - \gamma_1^3\right) + \frac{3A\left(A_x k^2 + Akk_x\right)}{k^3} \\
& \times \left[\frac{1}{6}\varepsilon Ak\gamma_1 - (\varepsilon km + (k - \omega))\gamma_2 - \frac{1}{2}\varepsilon Ak\gamma_3\right] \\
& = 0, \qquad (3.94)
\end{aligned}
$$

where use has been made of the periodicity of Θ_p.

In this way we obtain (3.82) and a set of first-order differential equations (i.e. (3.84), (3.91) and (3.94)) for k, ω, m and A. In the general case these equations are rather difficult to solve. Hence we are led to consider the approximate expansions of solutions for small ε. First we make some remarks. For $\varepsilon = 0$ the original equation (3.74) is linear and we have the dispersion relation (3.83), which is independent of A and h; hence the dependence of ω on A and h is a nonlinear effect; moreover, since we have $\gamma_1 = \gamma_3 = 0$, we shall have trivial solution from (3.91) and (3.94), if we put ε equal to zero. In order to take these facts into account, we introduce, instead of m, the quantity m_1 such that $h = \varepsilon m_1 + A\gamma_1$ is the mean wave height. Neglecting the terms of order ε^2, from (3.91) and (3.94) we then arrive at the following:

$$
\begin{aligned}
& \frac{\partial h}{\partial t} + \frac{\partial h}{\partial x} + \frac{1}{2}\varepsilon\gamma_2\frac{\partial A^2}{\partial x} = 0, \\
& \frac{\partial A^2}{\partial t} + \frac{\partial}{\partial x}\left[\left(1 - 12\pi^2 k^2\right) A^2\right] = 0,
\end{aligned}
\qquad (3.95)
$$

together with the dispersion relation (3.83) which can be written as:

$$
\omega_0 = k - 4\pi^2 k^3. \qquad (3.96)
$$

Here we write ω_0 instead of ω in order to emphasize the relation corresponding to the linear problem. The second equation of (3.95) is obtained by making use of (3.96); we observe that $\omega_0'(k) = 1 - 12\pi^2 k^2$. Finally, we introduce the second order functions $H_0(x,t)$ and $E(x,t)$ by means of

$$H_0 = \varepsilon h, \qquad E = \varepsilon^2 A^2,$$

in terms of which equations (3.95) may be written in the form:

$$\begin{aligned} &\frac{\partial H_0}{\partial t} + \frac{\partial}{\partial x}\left(H_0 + \frac{\gamma_2}{2} E\right) = 0, \\ &\frac{\partial E}{\partial t} + \frac{\partial}{\partial x}\left\{\omega_0'(k) E\right\} = 0. \end{aligned} \tag{3.97}$$

The first one of the above equations may be considered as an average form of the equation of conservation of mass; the second one expresses the conservation of the average energy of the wave train with a linear propagation group velocity $\omega_0'(k) = 1 - 12\pi^2 k^3$. It is possible to derive from the dispersion relation (3.82) an asymptotic expansion of ω in terms of k, H_0 and E for small ε. The latter together with (3.84) and (3.97) can then be utilized for determining ω as well as k, H_0 and E approximately.

CHAPTER IV - DEEP WATER SHIP HYDRODYNAMICS

In this chapter we describe the flow past a thin ship with uniform speed. With the help of Fourier transforms a solution of the linearized free surface problem is obtained. For the wave resistance the theory results in the classical '*Michell integral*' (1898).

The second part of this chapter is mainly concerned with the ships motion problem for a ship at constant forward speed. The potential function is written as a superposition of a steady and an unsteady part. A proper linearization shows that in the free surface condition for the unsteady part the undisturbed forward speed U plays a role. Application of a Laplace transform with respect to time results in expressions for the equations of motion, in all six degrees of freedom, of the convolution type as originally given by Ogilvie [7]. In principle ship forms are considered for which source distributions along the ship hull yield accurate approximation of the potential fields considered. For more general forms line integrals have to be taken into account as well, as will be indicated.

4.1 The Michell Theory

We consider a symmetrical ship moving with a constant speed U in the direction of the $\bar{x}$-axis of a right handed coordinate system $\bar{x},\bar{y},\bar{z}$ where the $\bar{y}$-axis is taken positive in downward direction. For convenience, we shall fix the coordinate system to the ship, or in other words, we consider flow past the ship with a uniform speed U in the negative $\bar{x}$-direction (Figure 4.1). We assume the equation of the hull to be given by $\bar{z} = \pm\bar{f}(\bar{x},\bar{y})$ where $\bar{f}$ is a smooth function defined on the projection of the ship hull on the $\bar{x},\bar{y}$-plane. The beam B of the ship is small compared to its length L.

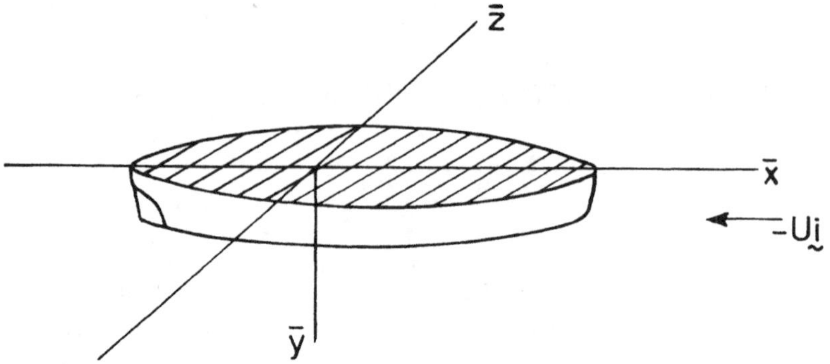

Figure 4.1

The perturbation caused by the ship is described by a velocity potential $\bar{\varphi}$ which satisfied Laplace equation together with the linearized free surface condition, obtained from (1.32) which now reads

$$U^2\bar{\varphi}_{\bar{x}\bar{x}} - g\bar{\varphi}_{\bar{y}} = 0, \quad \text{at } \bar{y} = 0 \tag{4.1}$$

and the boundary condition on the ship hull. This boundary condition states that the total normal velocity relative to the ship vanishes. According to the coordinate system used, we have an unperturbed speed $-U$ in the $\bar{x}$-direction and hence the velocity is $(-U + \bar{\varphi}_{\bar{x}}, \bar{\varphi}_{\bar{y}}, \bar{\varphi}_{\bar{z}})$. Since $(\pm\bar{f}_{\bar{x}}, \pm\bar{f}_{\bar{y}}, -1)$ are direction numbers of the normal to the ship hull, we obtain finally the condition:

$$-U\bar{f}_{\bar{x}} + \bar{\varphi}_{\bar{x}}\bar{f}_{\bar{x}} + \bar{\varphi}_{\bar{y}}\bar{f}_{\bar{y}} \mp \bar{\varphi}_{\bar{z}} = 0. \tag{4.2}$$

Now we introduce dimensionless coordinates $x = \bar{x}/L$, $y = \bar{y}/L$, $z = \bar{z}/L$ and the dimensionless velocity potential φ defined by

$$\bar{\varphi} = \varepsilon UL\varphi(x,y,z),$$

with $\varepsilon = B/L$. Clearly this does not affect the Laplace equation. However, the free surface condition (4.1) becomes

$$F^2\varphi_{xx} - \varphi_y = 0, \quad \text{for } y = 0, \tag{4.3}$$

where $F = U/(\sqrt{gL})$ is the Froude number. Furthermore, the condition (4.2) is transformed into the dimensionless form:

$$-f_x + \varepsilon\varphi_x f_x + \varepsilon\varphi_y f_y \mp \varphi_z = 0, \qquad \text{at } z = \pm Bf(x,y) \tag{4.4}$$

by setting $\bar{f}(x,y) = Bf(x,y)$. In the first order approximation, this condition reduces to

$$\varphi_z = \mp f_x \tag{4.5}$$

on the plane $z = 0$. Due to the symmetry of the flow we see that outside the ship hull $\varphi_z = 0$, for $z = 0$. We are thus led to the boundary value problem for φ defined by the Laplace equation:

$$\varphi_{xx} + \varphi_{yy} + \varphi_{zz} = 0, \tag{4.6}$$

together with the conditions:

$$\varphi \text{ finite,} \qquad \text{as } y \to \infty, \tag{4.7}$$

$$\varphi_z = \begin{Bmatrix} \mp f_x & \text{inside} \\ 0 & \text{outside} \end{Bmatrix} \begin{matrix} \text{the projection of the ship} \\ \text{hull on the plane } z = 0. \end{matrix}$$

There are two ways of solving this problem. The first one uses the Fourier transforms; the other one is by means of a distribution of sources on the center-plane. We shall use the first method. Applying the exponential Fourier transform with respect to x and the Fourier cosine transform with respect to z, we transform the Laplace equation for φ into the non-homogeneous equation:

$$\tilde{\varphi}_{yy} - (\alpha^2 + \beta^2)\tilde{\varphi} = \tilde{f}_x(\alpha,y), \tag{4.8}$$

with

$$\tilde{\varphi}(\alpha,y,\beta) = \int_{-\infty}^{\infty} e^{i\alpha x}dx \int_0^{\infty} \cos \beta z \, \varphi(x,y,z)dz$$

and

$$\tilde{f}_x(\alpha,y) = \int_0^\infty e^{i\alpha x} \varphi_z(x,y,0)\, dx.$$

Here we note that $f_x(\alpha,y)$ vanishes for y outside the projection of the ship hull on the plane $z = 0$. The surface condition at $y = 0$ is transformed in the same way. We have

$$-\alpha^2 F^2 \tilde{\varphi} - \tilde{\varphi}_y = 0, \qquad \text{at } y = 0. \tag{4.9}$$

Moreover, $\tilde{\varphi}$ remains finite as $y \to \infty$.

We construct the solution of the boundary value problem (4.8)-(4.9) by means of Green's function $G(y;\eta)$, i.e., a function which satisfies the conditions:

$$\begin{aligned}
&\text{a. } L[G] := G_{yy} - k^2 G = 0, && \text{for } y \neq \eta,\ k^2 = \alpha^2 + \beta^2, \\
&\text{b. } -\alpha^2 F^2 G - G_y = 0, && \text{at } y = 0, \\
&\text{c. } G \to 0, && \text{as } \bar{y} \to \infty, \qquad (4.10) \\
&\text{d. } \lim_{\varepsilon \to 0} \left(G_y(\eta + \varepsilon;\eta) - G_y(\eta - \varepsilon;\eta) \right) = -1, \\
&\text{e. } \lim_{\varepsilon \to 0} \left(G(\eta + \varepsilon;\eta) - G(\eta - \varepsilon;\eta) \right) = 0.
\end{aligned}$$

From (4.8) and (4.10)(a) it is easy to see that for $y \neq \eta$:

$$GL[\tilde{\varphi}] - \tilde{\varphi}L[G] = \frac{\partial}{\partial y}\left[\tilde{\varphi}_y G - G_y \tilde{\varphi}\right] = G(y;\eta)\, \tilde{f}_x(\alpha,y).$$

Integrating this expression over the interval of $0 \leqslant y < \infty$, we find

$$\begin{aligned}
\int_0^\infty G(y;\eta)\, \tilde{f}_x(\alpha,y)\, dy &= \int_0^\eta \frac{\partial}{\partial y}\left[\tilde{\varphi}_y G - G_y \tilde{\varphi}\right] dy \\
&\quad + \int_\eta^\infty \frac{\partial}{\partial y}\left[\tilde{\varphi}_y G - G_y \tilde{\varphi}\right] dy \\
&= \left[G\tilde{\varphi}_y - G_y \tilde{\varphi}\right] \Big|_{y=0}^{1} + \tilde{\varphi}(\eta)\, G_y(y;\eta) \Big|_{y=\eta-0}^{y=\eta+0}.
\end{aligned}$$

Because of the boundary condition (4.9) and (4.10)(b), the first term on the right cancels out, while the second term equals $-\tilde{\varphi}(\eta)$ by the jump condition (4.10)(d). Thus we obtain the representation

$$\tilde{\varphi}(\eta) = -\int_0^{\infty} \tilde{f}_x(\alpha,y)\; G(y;\eta)\; dy. \tag{4.11}$$

It now remains to derive the expression for the Green's function $G(y;\eta)$. By the conditions (4.10)(a)-(c) and (e) for G it is easy to see that

$$G(y;\eta) = \begin{cases} A\left(e^{ky} + \dfrac{k+\alpha^2F^2}{k-\alpha^2F^2}\, e^{-ky}\right), & 0 \leqslant y < \eta, \\[2ex] A\left(e^{k\eta-ky}\right)\left(e^{k\eta} + \dfrac{k+\alpha^2F^2}{k-\alpha^2F^2}\, e^{-k\eta}\right), & \eta < y < \infty. \end{cases}$$

Here the constant A has to be determined from the jump condition (4.10)(d). The final form for G can now be written as:

$$G(y;\eta) = \begin{cases} e^{-k\eta}\left(\dfrac{e^{ky} - e^{-ky}}{2k} + \dfrac{e^{-ky}}{k-\alpha^2F^2}\right), & 0 \leqslant y < \eta, \\[2ex] e^{-ky}\left(\dfrac{e^{k\eta} - e^{-k\eta}}{2k} + \dfrac{e^{-k\eta}}{k-\alpha^2F^2}\right), & \eta < y < \infty, \end{cases}$$

or simply:

$$G(y;\eta) = \frac{e^{-k|y-\eta|} - e^{-k(y+\eta)}}{2k} + \frac{e^{-k(y+\eta)}}{k-\alpha^2F^2}, \quad 0 \leqslant y,\eta < \infty. \tag{4.12}$$

To obtain the velocity potential φ we have to perform the inverse transformation. This gives

$$\varphi(x,y,z) = \frac{1}{2\pi^2}\int_{-\infty}^{\infty} e^{-i\alpha x}\,d\alpha \int_0^{\infty} \cos\beta z\,\tilde{\varphi}(\alpha,y,\beta)\,d\beta$$

$$= -\frac{1}{2\pi^2}\int_{-\infty}^{\infty} e^{-i\alpha x}\,d\alpha \int_0^{\infty} \cos\beta z d\beta \int_0^{\infty} \tilde{f}_x(\alpha;\eta)$$

$$\times \left[\frac{e^{-k|y-\eta|}}{2k} - \frac{e^{-k(y+\eta)}}{2k} + \frac{e^{-k(y+\eta)}}{k-\alpha^2 F^2}\right] d\eta, \qquad (4.13)$$

by making use of (4.11) and (4.12). Now we consider the first term in the above expressions. We have the first term

$$-\frac{1}{2\pi^2}\int_{-\infty}^{\infty} e^{-i\alpha x}\,d\alpha \int_0^{\infty} \cos\beta z d\beta \int_0^{\infty} \tilde{f}_x(\alpha;\eta)\,\frac{e^{-k|y-\eta|}}{2k}\,d\eta$$

$$= -\frac{1}{2\pi^2}\int_{-\infty}^{\infty} e^{-i\alpha x}\,d\alpha \int_0^{\infty} \cos\beta z d\beta \int_0^{\infty} \frac{e^{-k|y-\eta|}}{2k}\,d\eta$$

$$\times \int_0^{\infty} e^{i\alpha\tilde{x}}\,\varphi_z(\tilde{x},\eta,0)\,d\tilde{x},$$

which follows from the definition of $\tilde{f}_x$ in (4.8). Let us first consider the integration with respect to α and β. We see that

$$-\frac{1}{4\pi^2}\int_{-\infty}^{\infty}\int_{-\infty}^{\infty} \frac{1}{2k}\exp\left\{-i\alpha x + i\beta z + i\alpha\tilde{x} - k|y-\eta|\right\} d\alpha d\beta$$

$$= -\frac{1}{8\pi^2}\int_0^{\infty}\int_0^{2\pi} \exp\left\{-ik[(x-\tilde{x})\cos\vartheta - z\sin\vartheta]\right.$$

$$\left. - k|y-\eta|\right\} dk d\vartheta,$$

by using polar coordinates: $\alpha = k\cos\vartheta$, $\beta = k\sin\vartheta$ and $d\alpha d\beta = k dk d\vartheta$. Then from the formula,

$$\frac{1}{2\pi}\int_0^{2\pi} \exp\left\{-ik[(x-\tilde{x})\cos\vartheta - z\sin\vartheta]\right\} d\vartheta$$

$$= J_0\left(k\sqrt{(x-\tilde{x})^2 + z^2}\right),$$

it follows that

$$- \frac{1}{8\pi^2} \int_0^\infty \int_0^{2\pi} \exp \left\{ -ik[(x-\tilde{x}) \cos \vartheta - z \sin \vartheta] - k|y-\eta| \right\} dk d\vartheta$$

$$= - \frac{1}{4\pi} \int_0^\infty J_0 \left(k\sqrt{(x-\tilde{x})^2 + z^2} \right) e^{-k|y-\eta|} \, dk$$

$$= - \frac{1}{2\sqrt{\pi}} \left\{ (x-\tilde{x})^2 + z^2 + (y-\eta)^2 \right\}^{-1/2} .$$

The second term in (4.13) can be calculated similarly. Hence the first two terms in (4.13) give

$$- \frac{1}{4\pi} \int_{-\infty}^\infty d\tilde{x} \int_0^\infty \varphi_z(\tilde{x},\eta,0) \left\{ [(x-\tilde{x})^2 + z^2 + (y-\eta)^2]^{-1/2} - [(x-\tilde{x})^2 + z^2 + (y+\eta)^2]^{-1/2} \right\} d\eta, \tag{4.14}$$

the potential of source terms for the Laplace equation with surface condition $\varphi = 0$, at $y = 0$.

The last term in (4.13) will give the waves, resulting from the free surface condition,

$$- \frac{1}{8\pi^2} \int_{-\infty}^\infty e^{-i\alpha(x-\tilde{x})} \int_{-\infty}^\infty \frac{e^{i\beta z - k(y+\eta)}}{k - \alpha^2 F^2} \, d\alpha d\beta$$

$$= - \frac{1}{4\pi^2} \int_0^\infty \int_0^{2\pi} \frac{1}{1 - kF^2 \cos^2 \vartheta}$$

$$\times \exp \left\{ -ik[(x-\tilde{x}) \cos \vartheta - z \sin \vartheta] - k(y+\eta) \right\} dk d\beta . \tag{4.15}$$

Here, however, some care must be taken, since the integral is divergent. It is customary to avoid the difficulty by introduction of a small artificial damping term, which will eliminate the effect of the poles in the integrand with respect to the contour. We suppose that the downstream waves damp out. If the ship moves in the positive x-direction, the waves are carried along in the negative x-direction. We take account of this damping by giving U a small imaginary part, say $U - i\delta$, $\delta > 0$. This changes F to $F - i\delta$ and the pole in the integrand $k = 1/F^2 \cos^2 \vartheta$ will have a small positive imaginary part.

Figure 4.2

The contour moves under the pole. Then in the limit $\delta \to 0$, it gives the situation that we indent the contour by a half circle under the pole (see Figure 4.2 above).

Now collecting results from (4.14) and (4.15), we obtain

$$\varphi(x,y,z) = \mp \iint\limits_{S_0} G(x,y,z;\tilde{x},\tilde{y},0)\ f_x(\tilde{x},\tilde{y})\ d\tilde{x}d\tilde{y}, \tag{4.16}$$

where

$$G(x,y,z;\tilde{x},\tilde{y},0) = -\frac{1}{4\pi}\left[\frac{1}{r_1} - \frac{1}{r_2} + \int_0^{\infty}\int_0^{2\pi} \frac{1}{1 - kF^2\cos^2\vartheta}\right.$$

$$\left. \times \exp\left\{-ik[(x-\tilde{x})\cos\vartheta - z\sin\vartheta] - k(y+\tilde{y})\right\} dk d\vartheta\right]$$

$$r_1 = \left[(x-\tilde{x})^2 + z^2 + (y-\tilde{y})^2\right]^{1/2},$$

$$r_2 = \left[(x-\tilde{x})^2 + z^2 + (y+\tilde{y})^2\right]^{1/2},$$

and S_0 denotes the projection of ship hull on $z = 0$. The resistance of the ship can be found by integrating the pressure distribution over the hull. The pressure difference in the linearized approximation is defined by

$$p_+ - p_- = 2\rho U \varphi_x,$$

and the component in the x-direction of the total force R is defined by

$$R = 2\rho U \iint\limits_{S_0} \varphi_x(x,y,0)\ f_x(x,y)\ dxdy.$$

Working out this expression, we see that the terms $1/r_1$ and $1/r_2$ do not give any contribution to R, as they should not do according to the d'Alembert paradox which states that the total force acting on an object located in a potential flow is equal to zero.

After a rather tedious computation, the final result is

$$R = \frac{4g^2\rho}{\pi U^2} \int_0^{\pi/2} \sec^3 \vartheta \left[P^2(\vartheta) + Q^2(\vartheta) \right] d\vartheta, \tag{4.17}$$

where

$$P = \iint_{S_0} f_x(x,y)\, e^{Y \frac{\sec^2 \vartheta}{F^2}} \cos\left(\frac{x \sec \vartheta}{F^2}\right) dxdy,$$

$$Q = \iint_{S_0} f_x(x,y)\, e^{Y \frac{\sec^2 \vartheta}{F^2}} \sin\left(\frac{x \sec \vartheta}{F^2}\right) dxdy.$$

The result (4.17) may be put into a variety of different forms by change of variable and order of integration. If one denotes $\lambda = \sec \vartheta$, then it can be verified that

$$R = \frac{4g^2\rho}{\pi U^2} \iint_{S_0} dxdy \iint_{S_0} d\tilde{x}d\tilde{y}\; f_x(x,y)\, f_x(\tilde{x},\tilde{y}) \times M\left(\frac{x-\tilde{x}}{F^2}, \frac{y+\tilde{y}}{F^2}\right), \tag{4.18}$$

where

$$M(x,y) = \int_1^{\infty} \frac{\lambda^2}{\sqrt{\lambda^2 - 1}}\, e^{\lambda^2 y} \cos \lambda x d\lambda.$$

This expression, (4.18), for R is usually called '*Michell's integral*'.

4.2 Ship Motion

The motion of a body which has a forward speed U under the influence of incoming waves defines a boundary value problem for the velocity potential. We shall analyze this problem by introducing two coordinate systems: $\underset{\sim}{x} = (x_1,x_2,x_3)$ and $\underset{\sim}{x}' = (x_1',x_2',x_3')$; the former moves steadily with speed U in the direction of the positive x_1-axis while the latter is fixed to the ship. The two systems are supposed to coincide when the body is in its equilibrium. When the body is displaced from its equilibrium, the deflection of any point of the hull can be expressed as

$$\delta \underset{\sim}{\alpha}(\underset{\sim}{x},t) = \underset{\sim}{x} - \underset{\sim}{x}' = \delta \sum_{k=1}^{6} \underset{\sim}{\alpha}^{k}(\underset{\sim}{x},t), \tag{4.19}$$

where

$$\underset{\sim}{\alpha}^{k}(\underset{\sim}{x},t) = \begin{cases} \alpha_k(t)\, \underset{\sim}{i}_k , & k = 1,2,3 \\ \alpha_k(t)(\underset{\sim}{i}_{k-3} \wedge \underset{\sim}{x}), & k = 4,5,6. \end{cases}$$

Here δ is a small parameter; $\underset{\sim}{i}_k$ $(k = 1,2,3)$ are the unit vectors in the directions of x_k-axis $(k = 1,2,3)$; $\alpha_K(t)$ is the deflection in translational motion for $k = 1,2,3$, and for $k = 4,5,6$, $\alpha_k(t)$ represents rotation angles about x_{k-3}-axis. It is assumed that the α_k's are small enough so that the second order terms can be neglected.

We represent the potential by means of two small parameters ε and δ, the first one pertaining to the steady motion and the second one to the unsteady motion, and denote it by

$$\Phi_{total}(\underset{\sim}{x},t) = -Ux_1 + \varepsilon\Phi_0(\underset{\sim}{x}) + \delta\Phi_1(\underset{\sim}{x},t). \tag{4.20}$$

Here $-Ux_1 + \varepsilon\Phi_0(\underset{\sim}{x})$ is the potential for the steady flow past the ship fixed in its undisturbed position, that is, Φ_0 satisfies the linearized free surface condition:

$$U^2\Phi_{0x_1x_1} + g\Phi_{0x_3} = 0, \qquad \text{on } x_3 = 0, \tag{4.21}$$

$$\underset{\sim}{n}\cdot\underset{\sim}{V}_0(\underset{\sim}{x}) = 0 \qquad \text{on } S_0,$$

where

$$\underset{\sim}{V}_0(\underset{\sim}{x}) = \nabla\Big(-Ux_1 + \varepsilon\Phi_0(\underset{\sim}{x})\Big) = -U\underset{\sim}{i}_1 + \varepsilon\nabla\Phi_0,$$

and S_0 is the undisturbed position of the surface of the body (the undisturbed hull).

The free surface condition for the unsteady part of the motion $\delta\Phi_1$ is

$$\Phi_{1tt} - 2U\Phi_{1x_1t} + U^2\Phi_{1x_1x_1} + g\Phi_{1x_3} = 0, \text{ on } x_3 = 0, \tag{4.22}$$

but the condition on $\delta\Phi_1$ at the ship hull must be considered somewhat more carfully. The condition states that at the momentary position of the hull, the normal velocity of the fluid is equal to the normal velocity of the hull. The displacement from the rest position is $\delta\underset{\sim}{\alpha}$ and we take account of this effect by expanding Φ_{total} into a power series in $\delta\underset{\sim}{\alpha}$. To neglect second order and higher order terms of $\delta\underset{\sim}{\alpha}$, this gives the equation on the rest position S_0:

$$\underset{\sim}{n}\cdot\nabla\left\{\Phi_{total} + \delta\underset{\sim}{\alpha}\cdot\nabla\Phi_{total}\right\} = \delta\underset{\sim}{n}\cdot\frac{\partial\underset{\sim}{\alpha}}{\partial t}, \tag{4.23}$$

or

$$\underset{\sim}{n}\cdot\nabla\left\{-Ux_1 + \varepsilon\Phi_0 + \delta\Phi_1 + \delta\underset{\sim}{\alpha}\cdot V_0\right\} = \delta\underset{\sim}{n}\cdot\frac{\partial\underset{\sim}{\alpha}}{\partial t}. \tag{4.24}$$

Since on S_0, we have that $\underset{\sim}{n}\cdot\underset{\sim}{V}_0 = 0$ (cf. (4.21)), this reduces to

$$\underset{\sim}{n}\cdot\nabla\Phi_1 = \underset{\sim}{n}\cdot\left(\frac{\partial\underset{\sim}{\alpha}}{\partial t} - \nabla(\underset{\sim}{\alpha}\cdot\underset{\sim}{V}_0)\right). \tag{4.25}$$

In the general case, the function $\delta\Phi_1$ is the sum of three contributions, the incoming wave potential $\delta\Phi^{inc}$, the diffracted wave potential $\delta\Phi^d$, which is of the same order of magnitude as the incoming wave, and the forced wave $\delta\Phi^f$, which results from the unsteady motion of the body. The incoming wave $\delta\Phi^{inc}$ is given. For the diffracted wave we have the boundary condition that at the ship hull S_0 the normal velocity of the incoming wave must be cancelled, i.e.,

$$\underset{\sim}{n}\cdot\nabla\Phi^d = -\underset{\sim}{n}\cdot\nabla\Phi^{inc}. \tag{4.26}$$

For the forced wave we have from (4.25) and (4.26),

$$\underset{\sim}{n}\cdot\nabla\Phi^{f} = \underset{\sim}{n}\cdot\left(\frac{\partial\underset{\sim}{\alpha}}{\partial t} - \nabla(\underset{\sim}{\alpha}\cdot\underset{\sim}{V}_0)\right). \tag{4.27}$$

Both the diffracted wave and the forced wave have to satisfy a radiation condition.

By applying the Laplace transform to the problem for arbitrarily unsteady motions $\Phi(\underset{\sim}{x},t)$ we reduce it to the problem for $\varphi(\underset{\sim}{x},s)$:

$$\varphi(\underset{\sim}{x},s) = \int_0^\infty e^{-st}\,\Phi(\underset{\sim}{x},t)\,dt,$$

the Laplace transform of $\Phi(\underset{\sim}{x},t)$. Clearly the transformation does not affect the Laplace equation, nor the steady solution, but for the unsteady motion, the free surface condition (4.22) is transformed into

$$s^2\varphi_1 - 2Us\varphi_{1x_1} + U^2\varphi_{1x_1x_1} + g\varphi_{1x_3} = 0, \quad \text{on } x_3 = 0, \tag{4.28}$$

if we assume that $\Phi_{1t} = \Phi_1 = 0$ for $t = 0$. The boundary conditions (4.26), (4.27) at the hull S_0 are transformed into

$$\begin{aligned} \underset{\sim}{n}\cdot\nabla\varphi^{d} &= -\underset{\sim}{n}\cdot\nabla\varphi^{inc}, \text{ and} \\ \underset{\sim}{n}\cdot\nabla\varphi^{f} &= \underset{\sim}{n}\cdot\left(s\hat{\underset{\sim}{\alpha}} - \nabla(\hat{\underset{\sim}{\alpha}}\cdot\underset{\sim}{V}_0)\right), \end{aligned} \tag{4.29}$$

where $\hat{\underset{\sim}{\alpha}}$ is the Laplace transform of $\underset{\sim}{\alpha}(t)$. We assume that as $|\underset{\sim}{x}| \to \infty$, all the φ's except φ^{inc} vanish.

The above problem for φ will be simplified by making use of a suitable Green's function. For a moving ship, we define the Green's function $G(\underset{\sim}{x},\underset{\sim}{x}_p)$ as follows:

a. $G_{x_1x_1} + G_{x_2x_2} + G_{x_3x_3} = 0, \quad$ for $\underset{\sim}{x} \neq \underset{\sim}{x}_p$,

b. $G(x,x_p) = \dfrac{1}{|\underset{\sim}{x} - \underset{\sim}{x}_p|}$ + (regular harmonic function)

in a neighbourhood of $\underset{\sim}{x}_p$

$$\text{c. } s^2G + 2UsG_{x_1} + U^2G_{x_1x_1} + gG_{x_3} = 0, \quad \text{on } x_3 = 0, \quad (4.30)$$

$$\text{d. } \underset{\sim}{n}\cdot\nabla G = 0, \quad \text{on } S_0,$$

$$\text{e. } G \sim |\underset{\sim}{x}|^{-1}, \quad \text{as } |\underset{\sim}{x}| \to \infty.$$

We note that (4.30)(c) corresponds to the free surface condition for the reverse flow in comparison of (4.28).

We shall find an explicit expression for φ_1 by applying Green's theorem to the region Ω_R bounded by the free surface S^f, the ship hull S_0 and a large hemisphere S_R in the lower half-space (Figure 4.3). We have that

$$\iiint_{\Omega_R} \left(G\Delta\varphi_1 - \varphi_1\Delta G\right) dV = \left(\iint_{S_0} + \iint_{S^f} + \iint_{S_R}\right) \times \left(G\frac{\partial\varphi_1}{\partial n} - \varphi_1\frac{\partial G}{\partial n}\right) d\sigma - 4\pi\varphi_1(\underset{\sim}{x}_p), \tag{4.31}$$

where as usual $\frac{\partial}{\partial n}$ denotes the outer normal derivative, and $\varphi_1(\underset{\sim}{x}_p,s)$ has been abbreviated by $\varphi_1(\underset{\sim}{x}_p)$. The volume integral vanishes with understanding that Ω_R is replaced by $\Omega_R\backslash D(\underset{\sim}{x}_p)$ where $D(\underset{\sim}{x}_p)$ is a small neighbourhood of $\underset{\sim}{x}_p$, and that a limiting procedure is used. On S_0 from the condition (4.30)(d), we see that

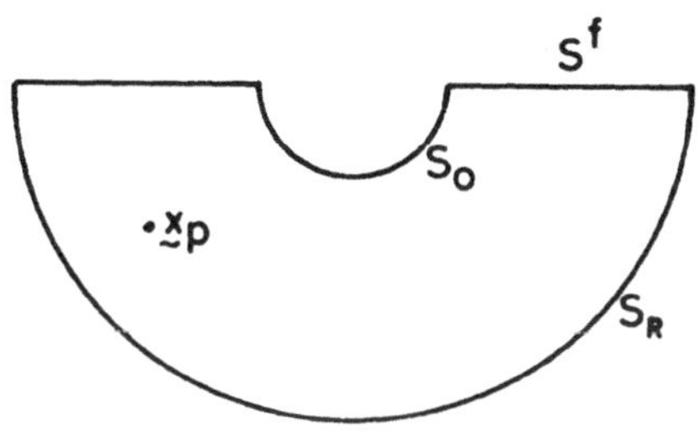

Figure 4.3

$$\iint_{S_0} \left(G \frac{\partial \varphi_1}{\partial n} - \varphi_1 \frac{\partial G}{\partial n} \right) d\sigma = \iint_{S_0} G \frac{\partial \varphi_1}{\partial n} d\sigma. \tag{4.32}$$

From the conditions at infinity for G and φ_1, it is easy to see that the integral on S_R goes to zero as $R \to \infty$. The integration over the free surface S^f, however, needs some consideration. Since the outer normal to S^f is in the direction of the positive x_3-axis, it follows that

$$g \frac{\partial \varphi_1}{\partial n} = g \frac{\partial \varphi_1}{\partial x_3} \quad \text{and} \quad g \frac{\partial G}{\partial n} = g \frac{\partial G}{\partial x_3}, \quad \text{on } x_3 = 0.$$

Consequently, we have from (4.28), (4.30)(c) that

$$\left.\begin{aligned} g \frac{\partial \varphi_1}{\partial n} &= -s^2\varphi_1 + 2Us \frac{\partial \varphi_1}{\partial x_1} - U^2 \frac{\partial^2 \varphi_1}{\partial x_1^2} \\ g \frac{\partial G}{\partial n} &= -s^2 G - 2Us \frac{\partial G}{\partial x_1} - U^2 \frac{\partial^2 G}{\partial x_1^2} \end{aligned}\right\} \text{on } x_3 = 0. \tag{4.33}$$

Hence, on the plane $x_3 = 0$ with the exception of the intersection with the body, we have

$$\iint_{S^f} \left(G \frac{\partial \varphi_1}{\partial n} - \varphi_1 \frac{\partial G}{\partial n} \right) dx_1 dx_2 = \frac{1}{g} \iint_{S^f} \left\{ 2sU \frac{\partial}{\partial x_1} (\varphi_1 G) - U^2 \left(G \frac{\partial^2 \varphi_1}{\partial x_1^2} - \varphi_1 \frac{\partial^2 G}{\partial x_1^2} \right) dx_1 dx_2, \right.$$

from which we can carry out the integration with respect to x_1. Then the left hand side of the above expression can be written in the form:

$$\frac{2sU}{g} \int (\varphi_1 G)_c \, dx_2 - \frac{U^2}{g} \int \left(G \frac{\partial \varphi_1}{\partial x_1} - \varphi_1 \frac{\partial G}{\partial x_1} \right)_c dx_2.$$

Here c stands for the corresponding x_1-boundary of S^f. At infinity, the integrated terms vanish and the remainder terms refer to an integral over the intersection of the body with the free surface. For slender ships the contribution of the line integral is small compared with the one of the surface integral and we then have from (4.31) and (4.32) the approximate result

$$\varphi_1(\underset{\sim}{x}_p) = \frac{1}{4\pi} \iint\limits_{S_0} G(\underset{\sim}{x},\underset{\sim}{x}_p) \frac{\partial\varphi_1}{\partial n} d\sigma. \tag{4.34}$$

Note that G is <u>not</u> symmetric in $\underset{\sim}{x}$ and $\underset{\sim}{x}_p$. In fact we have

$$G(\underset{\sim}{x},\underset{\sim}{x}_p) = G^*(\underset{\sim}{x}_p,\underset{\sim}{x}),$$

where G^* satisfies the free surface condition:

$$s^2G^* - 2UsG^*_{x_{p1}} + U^2G^*_{x_{p1}x_{p1}} + gG^*_{x_{p3}} = 0. \tag{4.35}$$

This can be easily verified by applying Green's theorem to the two functions $G(\underset{\sim}{x},\underset{\sim}{x}_p)$ and $G^*(\underset{\sim}{x},\underset{\sim}{x}_p)$ in the region considered above. The normal derivative $\partial\varphi_1/\partial n$ on the body consists of contributions from the incoming wave and that of each degree of freedom of the motion of the body. We have from (4.29)

$$\begin{aligned} \frac{\partial\varphi^d}{\partial n} &= -\frac{\partial\varphi^{inc}}{\partial n}; \\ \frac{\partial\varphi^f}{\partial n} &= \underset{\sim}{n}\cdot s\hat{\underset{\sim}{\alpha}} - \frac{\partial}{\partial n}(\hat{\underset{\sim}{\alpha}}\cdot\underset{\sim}{V}_0), \end{aligned} \tag{4.36}$$

where $\hat{\underset{\sim}{\alpha}}$ consists of the superposition

$$\hat{\underset{\sim}{\alpha}}(\underset{\sim}{x},s) = \sum_{k=1}^{6} \hat{\underset{\sim}{\alpha}}^k(\underset{\sim}{x},s), \tag{4.37}$$

with

$$\hat{\underset{\sim}{\alpha}}^k(\underset{\sim}{x},s) = \begin{cases} \hat{\alpha}_k(s)\, \underset{\sim}{i}_k, & k = 1,2,3 \\ \hat{\alpha}_k(s)(\underset{\sim}{i}_{k-3} \wedge \underset{\sim}{x}), & k = 4,5,6. \end{cases}$$

This gives six contributions to φ^f, defined by

$$\varphi^f = \sum_{k=1}^{6} \tilde{\alpha}_k \varphi_k^f, \tag{4.38}$$

with φ_k^f satisfying the boundary conditions:

$$\frac{\partial \varphi_k^f}{\partial n} = s\underset{\sim}{n}\cdot\underset{\sim}{i}_k - \frac{\partial}{\partial n}(\underset{\sim}{i}_k\cdot\underset{\sim}{V}_0), \qquad k = 1,2,3$$

$$\frac{\partial \varphi_k^f}{\partial n} = s\underset{\sim}{n}\cdot(\underset{\sim}{i}_{k-3}\wedge\underset{\sim}{x}) - \frac{\partial}{\partial n}\left\{(\underset{\sim}{i}_{k-3}\wedge\underset{\sim}{x})\cdot\underset{\sim}{V}_0\right\}, \quad k = 4,5,6. \tag{4.39}$$

From equation (4.36) and the above expressions for φ^f we can then express $\varphi_1(\underset{\sim}{x}_p)$ in (4.34) as

$$\begin{aligned} 4\pi\varphi_1(\underset{\sim}{x}_p) = &\sum_{k=1}^{3} \hat{\alpha}_k \iint_{S_0} G(\underset{\sim}{x},\underset{\sim}{x}_p)\left\{s\underset{\sim}{n}\cdot\underset{\sim}{i}_k - \frac{\partial}{\partial n}(\underset{\sim}{i}_k\cdot\underset{\sim}{V}_0)\right\} d\sigma \\ &+ \sum_{k=4}^{6} \hat{\alpha} \iint_{S_0} G(\underset{\sim}{x},\underset{\sim}{x}_p) \\ &\times\left\{s\underset{\sim}{n}\cdot(\underset{\sim}{i}_{k-3}\wedge\underset{\sim}{x}) - \frac{\partial}{\partial n}(\underset{\sim}{i}_{k-3}\wedge\underset{\sim}{x})\cdot\underset{\sim}{V}_0\right\} d\sigma \end{aligned} \tag{4.40}$$

For a body of arbitrary shape, it is not possible to give an explicit analytical formula for $G(\underset{\sim}{x},\underset{\sim}{x}_p)$. However, it should be noted that $G(\underset{\sim}{x},\underset{\sim}{x}_p)$ is not the Green's function for the original problem, but for the (Laplace) transformed problem and some properties of this function need to be considered.

From its construction $G(\underset{\sim}{x},\underset{\sim}{x}_p)$ has the form:

$$G(\underset{\sim}{x},\underset{\sim}{x}_p) = \frac{1}{|\underset{\sim}{x} - \underset{\sim}{x}_p|} + K_o(\underset{\sim}{x},\underset{\sim}{x}_p;s),$$

where $K_o(\underset{\sim}{x},\underset{\sim}{x}_p;s)$ has no singularities in the half-space under the body. The further decomposition now depends on the symmetry of the boundary conditions. We want to split from the functions φ_k^f a part which is independent of s, corresponding to the boundary value problem in an infinite fluid, and a part which results from the free surface, coming from the wave motion. Hence we study the boundary conditions from this point of view.

For the translational motion $\hat{\underset{\sim}{\alpha}}^1$, $\hat{\underset{\sim}{\alpha}}^2$ and the rotational motion $\hat{\underset{\sim}{\alpha}}^6$ in (4.38), which move on the x_3-plane, the boundary value problem for the complete body which results from the reflection of the wetted part with respect to the x_3-plane (Figure 4.4), the potential is symmetric with respect to this plane. This means that $\partial\varphi_k^f/\partial x_3 = 0$ for $k = 1$ and 2.
In this case we split off from the Green's function, the Green's function G_s for this boundary condition at the free surface,

$$G_s = \frac{1}{|\underset{\sim}{x} - \underset{\sim}{x}_p|} + \frac{1}{|\underset{\sim}{x} - \underset{\sim}{x}_p^*|} + g_s, \tag{4.41}$$

where $\partial g_s/\partial x_3 = 0$ at $x_3 = 0$ and $\partial G_s/\partial n = 0$ at S_0. Here $\underset{\sim}{x}_p^*$ is the mirror image of $\underset{\sim}{x}_p = (x_{p1}, x_{p2}, x_{p3})$, i.e., $\underset{\sim}{x}_p^* = (x_{p1}, x_{p2}, -x_{p3})$. This gives a uniquely defined decomposition of G, i.e.,

$$G(\underset{\sim}{x}, \underset{\sim}{x}_p) = G_s(\underset{\sim}{x}, \underset{\sim}{x}_p) + K_s(\underset{\sim}{x}, \underset{\sim}{x}_p; s). \tag{4.42}$$

For the translation motion $\hat{\underset{\sim}{\alpha}}^3$, as well as for the rotations $\hat{\underset{\sim}{\alpha}}^4$ and $\hat{\underset{\sim}{\alpha}}^5$, the potential is anti-symmetric with respect to the plane $x_3 = 0$. This gives the following condition that $\Phi_3^f = 0$ for $x_3 = 0$. The corresponding Green's function now is

$$G_a = \frac{1}{|\underset{\sim}{x} - \underset{\sim}{x}_p|} - \frac{1}{|\underset{\sim}{x} - \underset{\sim}{x}_p^*|} + g_a, \tag{4.43}$$

where $g_a = 0$ at $x_3 = 0$ and $\partial G_a/\partial n = 0$ at S_0. In this case the decomposition of G is

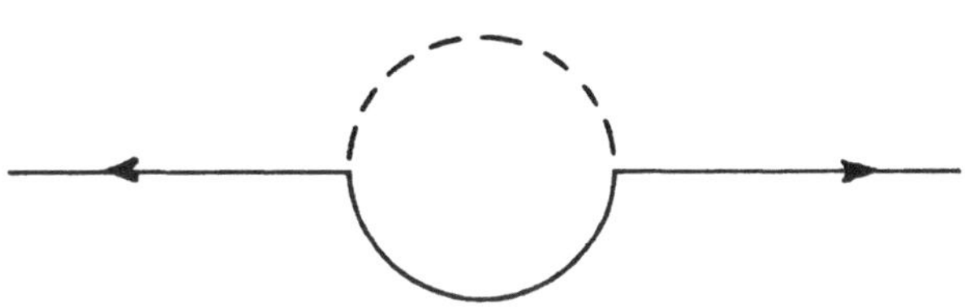

Figure 4.4

$$G(\underset{\sim}{x},\underset{\sim}{x}_p) = G_a(\underset{\sim}{x},\underset{\sim}{x}_p) + K_a(\underset{\sim}{x},\underset{\sim}{x}_p;s). \tag{4.44}$$

The reflection of the body does not introduce singularities, only if, as we assumed before, the intersection of the body with the free surface is perpendicular.

We are now in a position to decompose the explicit solution $\varphi_1(\underset{\sim}{x}_p)$ in (4.40) as

$$\begin{aligned}
\varphi_1(\underset{\sim}{x}_p) = & \sum_{k=1}^{2} \hat{\alpha}_k(s) \iint_{S_0} G_s(\underset{\sim}{x},\underset{\sim}{x}_p)\left\{s\underset{\sim}{n}\cdot\underset{\sim}{i}_k - \frac{\partial}{\partial n}(\underset{\sim}{i}_k\cdot\underset{\sim}{V}_0)\right\} d\sigma \\
& + \sum_{k=1}^{2} \hat{\alpha}_k(s) \iint_{S_0} K_s(\underset{\sim}{x},\underset{\sim}{x}_p;s)\left\{s\underset{\sim}{n}\cdot\underset{\sim}{i}_k - \frac{\partial}{\partial n}(\underset{\sim}{i}_k\cdot\underset{\sim}{V}_0)\right\} d\sigma \\
& + \hat{\alpha}_3(s) \iint_{S_0} G_a(\underset{\sim}{x},\underset{\sim}{x}_p)\left\{s\underset{\sim}{n}\cdot i_3 - \frac{\partial}{\partial n}(\underset{\sim}{i}_3\cdot\underset{\sim}{V}_0)\right\} d\sigma \\
& + \hat{\alpha}_3(s) \iint_{S_0} K_a(\underset{\sim}{x},\underset{\sim}{x}_p;s)\left\{s\underset{\sim}{n}\cdot\underset{\sim}{i}_3 - \frac{\partial}{\partial n}(\underset{\sim}{i}_3\cdot\underset{\sim}{V}_0)\right\} d\sigma \\
& + \sum_{k=4}^{5} \hat{\alpha}_k(s) \iint_{S_0} G_a(\underset{\sim}{x},\underset{\sim}{x}_p)\Big\{s\underset{\sim}{n}\cdot(\underset{\sim}{i}_{k-3}\wedge\underset{\sim}{x}) \\
& \qquad - \frac{\partial}{\partial n}\Big((\underset{\sim}{i}_{k-3}\wedge\underset{\sim}{x})\cdot\underset{\sim}{V}_0\Big)\Big\} d\sigma \\
& + \sum_{k=4}^{5} \hat{\alpha}_k(s) \iint_{S_0} K_a(\underset{\sim}{x},\underset{\sim}{x}_p;s)\Big\{s\underset{\sim}{n}\cdot(\underset{\sim}{i}_{k-3}\wedge\underset{\sim}{x}) \\
& \qquad - \frac{\partial}{\partial n}\Big((\underset{\sim}{i}_{k-3}\wedge\underset{\sim}{x})\cdot\underset{\sim}{V}_0\Big)\Big\} d\sigma \\
& + \hat{\alpha}_6(s) \iint_{S_0} G_s(\underset{\sim}{x},\underset{\sim}{x}_p)\Big\{s\underset{\sim}{n}\cdot(\underset{\sim}{i}_3\wedge\underset{\sim}{x}) \\
& \qquad - \frac{\partial}{\partial n}\Big((\underset{\sim}{i}_3\wedge\underset{\sim}{x})\cdot\underset{\sim}{V}_0\Big)\Big\} d\sigma \\
& + \hat{\alpha}_6(s) \iint_{S_0} K_s(\underset{\sim}{x},\underset{\sim}{x}_p;s)\Big\{s\underset{\sim}{n}\cdot(\underset{\sim}{i}_3\wedge\underset{\sim}{x}) \\
& \qquad - \frac{\partial}{\partial n}\Big((\underset{\sim}{i}_3\wedge\underset{\sim}{x})\cdot\underset{\sim}{V}_0\Big)\Big\} d\sigma.
\end{aligned} \tag{4.45}$$

Recall that $\varphi_1(\underset{\sim}{x}_p) = \varphi_1(\underset{\sim}{x}_p, s)$. It now remains to consider the inverse Laplace transform of φ_1 in (4.45). Since G_s and G_a do not contain the variable s, this means that the first parts give instantaneous contributions to the velocity potential, but the remainder parts will lead to convolution integrals over the past history of the motion. We introduce abbreviations for the integrals as follows:

$$\Phi_1(\underset{\sim}{x},t) = \sum_{k=1}^{6} \dot{\alpha}_k(t)\, \psi_{1k}(\underset{\sim}{x}) + \sum_{k=1}^{6} \alpha_k(t)\, \psi_{2k}(\underset{\sim}{x})$$

$$+ \sum_{k=1}^{6} \int_0^t \chi_{1k}(\underset{\sim}{x}, t-\tau)\, \dot{\alpha}_k(\tau)\, d\tau$$

$$+ \sum_{k=1}^{6} \int_0^t \chi_{2k}(\underset{\sim}{x}, t-\tau)\, \alpha_k(\tau)\, d\tau, \tag{4.46}$$

where

$$\psi_{1k}(\underset{\sim}{x}) = \iint_{S_0} G_s \underset{\sim}{n}\cdot\underset{\sim}{i}_k\, d\sigma; \qquad k = 1,2$$

$$\psi_{2k}(\underset{\sim}{x}) = \iint_{S_0} G_s \frac{\partial}{\partial n}(\underset{\sim}{i}_k\cdot\underset{\sim}{V}_0)\, d\sigma, \qquad k = 1,2$$

$$\psi_{13}(\underset{\sim}{x}) = \iint_{S_0} G_a \underset{\sim}{n}\cdot\underset{\sim}{i}_3\, d\sigma;$$

$$\psi_{23}(\underset{\sim}{x}) = \iint_{S_0} G_a \frac{\partial}{\partial n}(\underset{\sim}{i}_3\cdot\underset{\sim}{V}_0)\, d\sigma,$$

$$\psi_{1k}(\underset{\sim}{x}) = \iint_{S_0} G_a \underset{\sim}{n}\cdot(\underset{\sim}{i}_{k-3} \wedge \underset{\sim}{y})\, d\sigma; \qquad k = 4,5$$

$$\psi_{2k}(\underset{\sim}{x}) = \iint_{S_0} G_a \frac{\partial}{\partial n}\Big((\underset{\sim}{i}_{k-3} \wedge \underset{\sim}{y})\cdot\underset{\sim}{V}_0\Big)\, d\sigma, \qquad k = 4,5$$

$$\psi_{16}(\underset{\sim}{x}) = \iint_{S_0} G_s n\cdot(\underset{\sim}{i}_3 \wedge \underset{\sim}{y})\, d\sigma;$$

$$\psi_{26}(\underset{\sim}{x}) = \iint_{S_0} G_s \frac{\partial}{\partial n} \left((\underset{\sim}{i}_3 \wedge \underset{\sim}{y}) \cdot \underset{\sim}{V}_0 \right) d\sigma,$$

$$\chi_{1k}(\underset{\sim}{x},t) = \iint_{S_0} K_s(t) \underset{\sim}{n} \cdot \underset{\sim}{i}_k \, d\sigma; \qquad k = 1,2$$

$$\chi_{2k}(\underset{\sim}{x},t) = \iint_{S_0} K_s(t) \frac{\partial}{\partial n} (\underset{\sim}{i}_k \cdot \underset{\sim}{V}_0) \, d\sigma, \qquad k = 1,2$$

$$\chi_{13}(\underset{\sim}{x},t) = \iint_{S_0} K_a(t) \underset{\sim}{n} \cdot \underset{\sim}{i}_3 \, d\sigma;$$

$$\chi_{23}(\underset{\sim}{x},t) = \iint_{S_0} K_a(t) \frac{\partial}{\partial n} (\underset{\sim}{i}_3 \cdot \underset{\sim}{V}_0) \, d\sigma,$$

$$\chi_{1k}(\underset{\sim}{x},t) = \iint_{S_0} K_a(t) \underset{\sim}{n} \cdot (\underset{\sim}{i}_{k-3} \wedge \underset{\sim}{y}) \, d\sigma; \qquad k = 4,5$$

$$\chi_{2k}(\underset{\sim}{x},t) = \iint_{S_0} K_a(t) \frac{\partial}{\partial n} \left((\underset{\sim}{i}_{k-3} \wedge \underset{\sim}{y}) \cdot \underset{\sim}{V}_0 \right) d\sigma, \quad k = 4,5$$

$$\chi_{16}(\underset{\sim}{x},t) = \iint_{S_0} K_s(t) \underset{\sim}{n} \cdot (\underset{\sim}{i}_3 \wedge \underset{\sim}{y}) \, d\sigma;$$

$$\chi_{26}(\underset{\sim}{x},t) = \iint_{S_0} K_s(t) \frac{\partial}{\partial n} \left((\underset{\sim}{i}_3 \wedge \underset{\sim}{y}) \cdot \underset{\sim}{V}_0 \right) d\sigma. \qquad (4.47)$$

Here $K_s(t)$ and $K_a(t)$ denote the Laplace inverse transforms of $K_s(\underset{\sim}{y},\underset{\sim}{x};s)$ and $K_a(\underset{\sim}{y},\underset{\sim}{x};s)$, respectively.

4.3 Forces and Moments

The forces exerted on the moving body are found from the pressure. For simplicity, we restrict the discussion to a linearized version in ε as well as δ (cf. (4.31)), and omit the terms with $\varepsilon\delta$, which in many cases cannot be neglected. The pressure p is given by the Bernoulli's equation (1.13). To the first order in ε and δ, it can be approximated by

$$p = \rho\left[-gx_3 + U\frac{\partial\Phi_0}{\partial x_1} - \delta\left(\frac{\partial\Phi_1}{\partial t} - U\frac{\partial\Phi_1}{\partial x_1}\right)\right]. \tag{4.48}$$

The pressure, however, must be evaluated at the actual location of the body, which essentially is unknown. This inconvenience is overcome by expressing the pressure at S_0, the undisturbed position of the body, in terms of the pressure at the actual surface and the difference, resulting from a displacement S to S_0.

The motion is described by a displacement vector:

$$\delta\underset{\sim}{\xi}(t) = \delta\sum_{k=1}^{3}\alpha_k(t)\,\underset{\sim}{i}_k,$$

and a rotation vector:

$$\delta\underset{\sim}{\vartheta}(t) = \delta\sum_{k=4}^{6}\alpha_k(t)\,\underset{\sim}{i}_{k-3}.$$

Then the displacement from the equilibrium is $\underset{\sim}{\xi} + \underset{\sim}{\vartheta}\wedge\underset{\sim}{x}$ and the pressure at a point on S is expressed as

$$p\Big|_S = p\Big|_{S_0} + \delta(\underset{\sim}{\xi} + \underset{\sim}{\vartheta}\wedge\underset{\sim}{x})\cdot\nabla p\Big|_{S_0}. \tag{4.49}$$

The force on the body, denoted by $\underset{\sim}{X}$, is found by integration of the pressure and its components are given by

$$X_j = \iint_S p(\underset{\sim}{n}\cdot\underset{\sim}{i}_j)\,d\sigma, \qquad j = 1,2,3. \tag{4.50}$$

The factors $\underset{\sim}{n}\cdot\underset{\sim}{i}_j$ change during the motion, but the factors $\underset{\sim}{n}\cdot\underset{\sim}{i}'_j$ which are related to the x'_j-axis fixed to the body do not change. Hence it is more appropriate to express the force components in the moving frame (x'_1,x'_2,x'_3), such that

$$X'_j = \iint_S p(\underset{\sim}{n}\cdot\underset{\sim}{i}'_j)\,d\sigma, \qquad j = 1,2,3. \tag{4.51}$$

The corresponding force $\underset{\sim}{X}'$ is then related to that of the fixed frame (x_1,x_2,x_3) by

$$\underset{\sim}{X} = \underset{\sim}{X}' + \delta(\underset{\sim}{\vartheta} \wedge \underset{\sim}{X}'). \tag{4.52}$$

We now transform the expansion for X_j' into a form with the integration over S_0. This transformation has to take account of two effects: the displacement effects on p and the effect that the intersection with the free surface of S, the actual position of the body, is different from the intersection of S_0 with the free surface. This gives

$$X_j' = \iint_{S_0} (\underset{\sim}{n}\cdot\underset{\sim}{i}_j)\left\{p + \delta(\underset{\sim}{\xi} + \underset{\sim}{\vartheta} \wedge \underset{\sim}{x})\cdot\nabla p\right\} d\sigma_0$$
$$+ \int_L (\underset{\sim}{n}\cdot\underset{\sim}{i}_j') \, ds \int_0^{\zeta-\delta(\underset{\sim}{\xi}+\underset{\sim}{\vartheta}\wedge\underset{\sim}{x})\cdot\underset{\sim}{i}_3} dx_3$$
$$\times\left\{p + \delta(\underset{\sim}{\xi} + (\underset{\sim}{\vartheta} \wedge \underset{\sim}{x}))\cdot\nabla p\right\}.$$

Here the second integral is the contribution of the last effect; L denotes the intersection of the body with the free surface; the height of the wetted part of the ship is equal to the actual free surface height ζ plus the actual position of the free surface intersection in the rest position. Recall that the free surface height due to the unsteady motion is

$$\zeta = -\frac{\delta}{g}\left[\frac{\partial\Phi_1}{\partial t} - U\frac{\partial\Phi_1}{\partial x_1}\right],$$

(cf. (1.33)). Using this expression for ζ, with the pressure gradient from (4.48),

$$\nabla p = \rho\left[-g\underset{\sim}{i}_3 + \varepsilon U\frac{\partial}{\partial x_1}\nabla\Phi_0 - \delta\left(\frac{\partial\nabla\Phi_1}{\partial t} - U\frac{\partial\nabla\Phi_1}{\partial x_1}\right)\right],$$

it is easy to see that this integral does not give a contribution to the first order approximation. Thus, the result for X_j' is, in the first order approximation:

$$X_j' = \rho\iint_{S_0} (\underset{\sim}{n}\cdot\underset{\sim}{i}_j)\left\{-gx_3 - g\delta(\xi_3 + x_2\vartheta_1 - x_1\vartheta_2)\right.$$
$$\left.-\delta\left(\frac{\partial\Phi_1}{\partial t} - U\frac{\partial\Phi_1}{\partial x_1}\right) + \varepsilon U\frac{\partial\Phi_0}{\partial x_1}\right\} d\sigma. \tag{4.53}$$

For the moments X_{j+3}, we find in a similar way that

$$X'_{j+3} = \rho \iint_{S_0} (\underset{\sim}{n}\cdot\underset{\sim}{i}_j \wedge \underset{\sim}{x})\Big\{-gx_3 - g\delta(\xi_3 + x_2\vartheta_1 - x_1\vartheta_2) -\delta \left(\frac{\partial\Phi_1}{\partial t} - U\frac{\partial\Phi_1}{\partial x_1}\right) + \varepsilon U \frac{\partial\Phi_0}{\partial x_1}\Big\}\, d\sigma. \tag{4.54}$$

In equation (4.53), obviously the first term gives in the equilibrium position, the hydrostatic force: $X'_1 = 0$, $X'_2 = 0$, $X'_3 = \rho g V$, where V stands for the volume of the body below the free surface; the second term gives the hydrostatic disturbing force; the third term gives the force from the unsteady motion and the last term, the force from the steady motion. Because of the linearization, no interaction terms occur.

We are mainly interested in the forces due to the unsteady motion, which we will relate to the contributions of its components. Putting

$$\begin{aligned} F'_j &= -\rho g\delta \iint_{S_0} (\underset{\sim}{n}\cdot\underset{\sim}{i}_j)\left(\frac{\partial\Phi_1}{\partial t} - U\frac{\partial\Phi_1}{\partial x_1}\right) d\sigma, \\ F'_{j+3} &= -\rho g\delta \iint_{S_0} \Big(\underset{\sim}{n}\cdot(\underset{\sim}{i}_j \wedge \underset{\sim}{x})\Big)\left(\frac{\partial\Phi_1}{\partial t} - U\frac{\partial\Phi_1}{\partial x_1}\right) d\sigma, \end{aligned} \tag{4.55}$$

and using equation (4.46), i.e.,

$$\begin{aligned} \Phi_1 = {} & \sum_{k=1}^{6} \dot{\alpha}_k(t)\, \psi_{1k}(\underset{\sim}{x}) + \sum_{k=1}^{6} \alpha_k(t)\, \psi_{2k}(\underset{\sim}{x}) \\ & + \sum_{k=1}^{6} \int_0^t \chi_{1k}(x,t-\tau)\, \dot{\alpha}_k(\tau)\, d\tau \\ & + \sum_{k=1}^{6} \int_0^t \chi_{2k}(x,t-\tau)\, \alpha_k(\tau)\, d\tau, \end{aligned}$$

we can obtain expressions for F'_j and F'_{j+3} in terms of $\alpha_k(t)$, ψ_{ij} and etcetera. These expressions can be simplified if one is only interested in the first order approximations. In this approximation, if we let $\underset{\sim}{V}_0 = -U\underset{\sim}{i}_1$, (cf. (4.21)) then

$$\frac{\partial}{\partial n}\,(\underset{\sim}{i}_k\cdot\underset{\sim}{V}_0) = 0, \qquad k = 1,2,3$$

and

$$\frac{\partial}{\partial n}\left[(\underset{\sim}{i}_{k-3}\wedge\underset{\sim}{x})\cdot\underset{\sim}{V}_0\right] = -U(\underset{\sim}{i}_1\wedge\underset{\sim}{i}_{k-3})\cdot\underset{\sim}{n}, \qquad k = 4,5,6.$$

Hence, in equation (4.55), terms such as ψ_{2k} and χ_{2k} are equal to zero except that

$$\psi_{25}(\underset{\sim}{x}) = -U\iint_{S_0} G_a(\underset{\sim}{i}_3\cdot\underset{\sim}{n})\,d\sigma;$$

$$\psi_{26}(\underset{\sim}{x}) = -U\iint_{S_0} G_s(\underset{\sim}{i}_3\cdot\underset{\sim}{n})\,d\sigma,$$

$$\chi_{25}(\underset{\sim}{x}) = -U\iint_{S_0} K_a(t)(\underset{\sim}{i}_3\cdot\underset{\sim}{n})\,d\sigma;$$

$$\chi_{26}(\underset{\sim}{x}) = -U\iint_{S_0} K_s(t)(\underset{\sim}{i}_3\cdot\underset{\sim}{n})\,d\sigma.$$

Now we can write the expression for the force F_j' in the form:

$$\begin{aligned}
F_j' = &-\rho g\delta\sum_{k=1}^{6}\ddot{\alpha}_k(t)\iint_{S_0}(\underset{\sim}{n}\cdot\underset{\sim}{i}_j)\,\psi_{1k}(x)\,d\sigma\\
&+\rho g\delta U\sum_{k=1}^{6}\dot{\alpha}_k(t)\iint_{S_0}(\underset{\sim}{n}\cdot\underset{\sim}{i}_j)\,\frac{\partial\psi_{1k}}{\partial x_1}(\underset{\sim}{x})\,d\sigma\\
&-\rho g\delta\sum_{k=1}^{6}\dot{\alpha}_k(t)\iint_{S_0}(\underset{\sim}{n}\cdot\underset{\sim}{i}_j)\,\psi_{2k}(\underset{\sim}{x})\,d\sigma\\
&+\rho g\delta U\sum_{k=1}^{6}\alpha_k(t)\iint_{S_0}(\underset{\sim}{n}\cdot\underset{\sim}{i}_j)\,\frac{\partial\psi_{2k}}{\partial x_1}(\underset{\sim}{x})\,d\sigma\\
&+\rho g\delta U\sum_{k=1}^{6}\int_0^t\dot{\alpha}_k(\tau)\,d\tau\iint_{S_0}(\underset{\sim}{n}\cdot\underset{\sim}{i}_j)\,\frac{\partial\chi_{ik}}{\partial x_1}(\underset{\sim}{x},t-\tau)\,d\sigma\\
&-\rho g\delta\sum_{k=1}^{6}\int_0^t\dot{\alpha}_k(\tau)\,d\tau\iint_{S_0}(\underset{\sim}{n}\cdot\underset{\sim}{i}_j)\,\frac{\partial\chi_{1k}}{\partial t}(\underset{\sim}{x},t-\tau)\,d\sigma
\end{aligned}$$

$$+ \rho g \delta U \sum_{k=1}^{6} \int_0^t \alpha_k(\tau)\, d\tau \iint_{S_0} (\underset{\sim}{n} \cdot \underset{\sim}{i}_j) \frac{\partial \chi_{2k}}{\partial x_1} (\underset{\sim}{x}, t-\tau)\, d\sigma$$

$$- \rho g \delta \sum_{k=1}^{6} \int_0^t \alpha_k(\tau)\, d\tau \iint_{S_0} (\underset{\sim}{n} \cdot \underset{\sim}{i}_j) \frac{\partial \chi_{2k}}{\partial t} (\underset{\sim}{x}, t-\tau)\, d\sigma . \tag{4.56}$$

A similar expression holds for the moments:

$$F'_{j+3} = -\rho g \delta \sum_{k=1}^{6} \ddot{\alpha}_k(t) \iint_{S_0} \underset{\sim}{n} \cdot (\underset{\sim}{i}_{j+3} \wedge \underset{\sim}{x})\, \psi_{1k}(\underset{\sim}{x})\, d\sigma + \ldots \tag{4.57}$$

We omit the detail.

It is customary in expressions of this kind to denote the term which is proportional to $\ddot{\alpha}_k$ by *added (virtual) mass* and the term proportional to $\dot{\alpha}_k$ by *virtual damping*. It appears that these terms have no relation to the waves, since they belong to a boundary value problem in an infinite fluid. The damping terms come from the translational motion and the waves which are left behind. The unsteady motion gives rise to the convolution integrals which relate the previous history of the motion. They do not contribute to the virtual mass.

CHAPTER V - SHALLOW WATER SHIP HYDRODYNAMICS

In this chapter, we consider slender ships in shallow water. We shall discuss three related topics. We are, among others, interested in the influence of the bottom on the vertical position of the ship. It turns out that a ship in shallow water experiences a certain sinkage and trim due to the bottom effect. This is of importance if one wishes to determine the required depth of harbours in such a way that a ship may enter safely; it is also of i portance to know wall effects for ships travelling in a shallow channel. Here we wish to know sinkage and trim but also the force and moment due to interaction with the walls of the channel; the third topic we shall consider is the interaction between ships.

These topics have been thoroughly investigated by E.O. Tuck [10], R.F. Beck [1] and R.W. Yeung [15,16]. The first two are steady flow problems, while the third one is unsteady due to the difference in forward speed of the two ships. The method we use to solve these problems is related to the well-known method of matched asymptotic expansions. To get insight in this method, we apply it to the unsteady flow around a two-dimensional airfoil. The incoming flow is in principle nearly parallel to the airfoil while the fluid domain is extended toward infinity. Thereafter, we treat the case of a cross-current flow around a slender body wedged between two parallel plates. The notion 'blockage' will be introduced in this way. After these preliminary investigations, we shall pay attention to the three problems mentioned above.

5.1 Thin Airfoil Theory

We consider the two-dimensional thin airfoil moving with a constant translation velocity $U\underset{\sim}{i}$ in an incompressible medium (see Figure 5.1). With regard to a system of reference, fixed to the airfoil, the x-axis extending along the surface (approximately), the resulting motion of the fluid can be separated into an unperturbed flow with velocity U in the direction of the positive x-axis and a perturbation flow $\underset{\sim}{v} = (u,v)$ which originates from the airfoil.

The governing equations (1.1) and (1.2) now read:

$$u_x + v_y = 0, \tag{5.1}$$

$$\begin{aligned} u_t + (U + u)u_x + vu_y &= -\frac{1}{\rho} p_x, \\ v_t + (U + u)v_x + vv_y &= -\frac{1}{\rho} p_y, \end{aligned} \tag{5.2}$$

where p and ρ are pressure and density of the fluid. Because $|\underset{\sim}{v}| = \sqrt{u^2 + v^2} << U$, we may linearize the equation of motion (5.2), so that

$$\begin{aligned} u_t + Uu_x &= -\frac{1}{\rho} p_x, \\ v_t + Uv_x &= -\frac{1}{\rho} p_y. \end{aligned} \tag{5.3}$$

We now eliminate p from these two equations and introduce the vorticity vector

$$\underset{\sim}{\gamma} := -\left(\frac{\partial v}{\partial x} - \frac{\partial u}{\partial y}\right) \underset{\sim}{k} = \gamma \underset{\sim}{k},$$

where $\underset{\sim}{k}$ is the unit vector in the direction normal to the x,y-plane. This leads to the equation for the scalar function $\gamma = \gamma(x,y,t)$:

$$\gamma_t + U\gamma_x = 0, \tag{5.4}$$

from which we see that γ must have the form

$$\gamma = \gamma(x - Ut, y), \tag{5.5}$$

i.e., the vorticity is in linearized approximation carried along the undisturbed flow, in general along the stream lines. In the region where $\gamma = 0$, the velocity field has a velocity potential $\tilde{\Phi}$ such that

$$\tilde{\underset{\sim}{v}} = \text{grad}\, \tilde{\Phi},$$

and if the total field is given by

$$\tilde{\Phi} = Ux + \Phi,$$

clearly then the velocity potential Φ satisfies the two-dimensional Laplace equation:

$$\Phi_{xx} + \Phi_{yy} = 0. \tag{5.6}$$

Moreover, the pressure p and the potential Φ are related by the linearized Bernoulli equation:

$$\Phi_t + U\Phi_x = - \frac{p}{\rho} + f(t), \tag{5.7}$$

where f(t) is an arbitrary function of t.

To formulate the boundary value problem, we now consider a thin airfoil with the upper and lower surface given by

$$y = f(x,t) \pm g(x,t),$$

where the camber function $f(x,t) = \varepsilon F(x,t)$ and the thickness

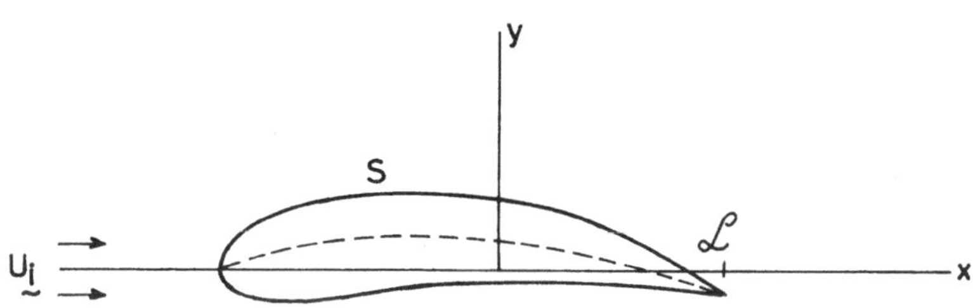

Figure 5.1

function $g(x,t) = \varepsilon G(x,t)$. Here ε is the small slenderness parameter. We denote the boundary of the airfoil by S. Then, the boundary value problem is defined by (5.6) in the exterior region to the airfoil together with the boundary conditions (cf. (1.9)):

$$\frac{\partial \Phi}{\partial n} = - \frac{\partial}{\partial n} (Ux), \tag{5.8}$$

or

$$\frac{\partial \Phi}{\partial y} = \varepsilon \left\{ (F \pm G)_t + \left(\frac{\partial \Phi}{\partial x} + U \right) (F \pm G)_x \right\}, \text{ on } S,$$

and the condition at infinity

$$\nabla \Phi \to \underset{\sim}{0}, \qquad \text{as } |\underset{\sim}{x}| \to \infty. \tag{5.9}$$

First we consider the field in the vicinity of the airfoil where $y = O(\varepsilon)$. This can be done by a proper stretching of the coordinate

$$y = \varepsilon Y.$$

In this inner region, we denote the potential by Φ^i, the inner potential while outside the region by Φ^0, the outer potential. We assume that Φ^i has a regular expansion in ε:

$$\Phi^i \sim \sum_{k=1}^{\infty} \varepsilon^k \Phi^i_k(x,Y,t). \tag{5.10}$$

In terms of the stretched variable Y, the Laplace equation (5.6) in the near field becomes

$$\Phi^i_{YY} + \varepsilon^2 \Phi^i_{xx} = 0$$

into which by substituting (5.8), and equating like power of ε, we obtain:

$$\Phi^i_{kYY} = \begin{cases} 0, & k = 1,2 \\ -\Phi^i_{k-2xx}, & k = 3,4,\ldots \end{cases} \tag{5.11}$$

The boundary condition (5.8) on the surface $Y = F \pm G$ becomes

$$\varepsilon\Phi^i_{1Y} + \varepsilon^2\Phi^i_{2Y} + \ldots = \varepsilon^2 \left\{(F_t + UF_x) \pm (G_t + UG_x)\right\} + \ldots.. \tag{5.12}$$

Hence from (5.11) and (5.12) it is easy to see that for $Y \gtrless 0$:

$$\Phi^i_1(x,Y,t) = A^{\pm}_1(x,t),$$
$$\Phi^i_2(x,Y,t) = A^{\pm}_2(x,t) + \left\{(F_t + UF_x) \pm (G_t + UG_x)\right\} Y, \tag{5.13}$$

where $A^{\pm}_k$ are unknowns. To determine $A^{\pm}_k$ we now use the matching principle in singular perturbation theory. We first write $y = \varepsilon Y$. This leads to

$$\Phi^i(x,y,t,\varepsilon) = \varepsilon \left[A^{\pm}_1(x,t) + \left\{(F_t + UF_x) \pm (G_t + UG_x)\right\}y\right] + \varepsilon^2 A^{\pm}_2(x,t), \tag{5.14}$$

which is supposed to be matched with the outer potential Φ^0 in common region where both inner and outer solutions are valid.

The outer potential Φ^0 has a regular perturbation expansion in ε of the form:

$$\Phi^0(x,y,t,\varepsilon) \sim \sum_{k=1}^{\infty} \varepsilon^k \Phi^0_k(x,y,t). \tag{5.15}$$

One term of the outer potential in terms of inner variable then reads:

$$\Phi^0(x,y,t,\varepsilon) = \varepsilon\Phi^0_1(x,\varepsilon Y,t)$$
$$= \varepsilon\Phi^0_1(x,0,t) + \varepsilon^2 Y\Phi^0_{1y}(x,0,t) + \ldots \tag{5.16}$$

The matching rule now requires that one term of (5.14) equals two terms of (5.16) which yields

$$\Phi_1^0(x,\pm 0,t) = A_1^{\pm}(x,t)$$

and (5.17)

$$\Phi_{1y}^0(x,\pm 0,t) = (F_t + UF_x) \pm (G_t + UG_x), \qquad \text{on } \mathcal{L},$$

where $\mathcal{L}$ is the projection of the airfoil on the x-axis. This last relation is well-known in linearized airfoil theory. Hence $A_1^{\pm}$ is completely determined if one has Φ_1^0.

We notice that the pressure is continuous on the x-axis outside $\mathcal{L}$ and has jump discontinuities along $\mathcal{L}$ because of potential jump $[\Phi](x,t)$ defined by $[\Phi](x,t) := \Phi(x,+0,t) - \Phi(x,-0,t)$. The fact that the pressure jump $[p] = 0$ outside $\mathcal{L}$ leads to

$$[\Phi]_t + U[\Phi]_x = 0,$$

for the potential jump $[\Phi]$ which implies that $[\Phi]$ must be a function of the variable $x - Ut$, say:

$$[\Phi](x,t) = f(x - Ut). \tag{5.18}$$

This means that $[\Phi]$ propagates along lines $y = 0$. In front of the airfoil, for $x = -\infty$ we have $[\Phi] = 0$. In the wake behind the foil the value of $[\Phi]$ differs from zero. If at the trailing edge x_T, $[\Phi]$ is known to be

$$[\Phi] = h(x_T,t),$$

then we have from (5.18):

$$[\Phi](x,t) = h\left(x_T, t - \frac{x - x_T}{U}\right). \tag{5.19}$$

Only if $[\Phi]$ is specified in the wake, then the boundary value problem for the potential Φ^0 may be stated properly as we will see.

The formulation of the boundary condition (5.17) suggests a splitting of the outer potential in an even and an odd part:

$$\Phi^0(x,y,t) = \varphi_e + \varphi_o \tag{5.20}$$

where
$$\varphi_e(x,y,t) = \varphi_e(x,-y,t), \text{ and}$$
$$\varphi_o(x,y,t) = -\varphi_o(x,-y,t),$$

from which we obtain odd and even derivatives, respectively:

$$\begin{aligned} \frac{\partial\varphi_e}{\partial y}(x,y,t) &= -\frac{\partial\varphi_e}{\partial y}(x,-y,t), \\ \frac{\partial\varphi_o}{\partial y}(x,y,t) &= \frac{\partial\varphi_o}{\partial y}(x,-y,t). \end{aligned} \tag{5.21}$$

We now consider for $y \geqslant 0$,

$$\begin{aligned} \Phi^0(x,y,t) &= \varphi_e(x,y,t) + \varphi_o(x,y,t), \\ \Phi^0(x,-y,t) &= \varphi_e(x,y,t) - \varphi_o(x,y,t). \end{aligned} \tag{5.22}$$

The matching condition (5.17) implies

$$\left.\begin{aligned} \frac{\partial\varphi_e}{\partial y} &= \left(\frac{\partial}{\partial t} + U\frac{\partial}{\partial x}\right) G \\ \frac{\partial\varphi_o}{\partial y} &= \left(\frac{\partial}{\partial t} + U\frac{\partial}{\partial x}\right) F \end{aligned}\right\} \text{ at } y = 0, \tag{5.23}$$

Across the x-axis outside $\mathcal{L}$, φ_e is continuous but φ_o can be discontinuous with

$$\varphi_o(x_1,+0,t) = -\varphi_o(x_1,-0,t). \tag{5.24}$$

The discontinuity in the wake for the velocity potential Φ^0 gives a condition for the odd part φ_o. Hence the boundary value problem for Φ^0 splits up in two problems, one pertaining to the thickness distribution for the function φ_e, the other one pertaining to the camber distribution for the function φ_o.

To solve the two boundary value problems for φ_e and φ_o it is convenient to use the fact that both functions are solutions of the two-dimensional Laplace equation. Hence the use of the analysis of complex functions in the $z = x + iy$ plane is desirable. In particular, we introduce:

$$\chi(z,t) = \varphi + i\psi,$$

with ψ being the stream function and φ the potential function. Here $\chi(z,t)$ is an analytic function in the z-plane except on a part of the x-axis. Because of the structure of the boundary conditions on the x-axis, it is very convenient to consider the complex velocity

$$W(z,t) := \frac{d\chi}{dz} = u - iv. \tag{5.25}$$

We determine a function $W_e(z,t)$ which is analytic in the complex z-plane except on the part $\mathcal{L}$ of the x-axis. We take the leading edge at $x = -1$ and the trailing edge at $x = 1$. On the segment $-1 \leqslant x \leqslant 1$, $y = 0$, we have the relation

$$W_e^+ - W_e^- = -2i \left(\frac{\partial}{\partial t} + U \frac{\partial}{\partial x}\right) G,$$

where $W^{\pm}$ denotes $W(x \pm i0,t)$, while for $|x| > 1$, $y = 0$,

$$W_e^+ - W_e^- = 0.$$

The solution $W_e(z,t)$ can be represented in the form:

$$W_e(z,t) = -\frac{1}{\pi} \int_{-1}^{+1} \frac{\left(\frac{\partial}{\partial t} + U \frac{\partial}{\partial \xi}\right) G}{\xi - z} \, d\xi, \tag{5.26}$$

because the function $W_e(z,t)$ has to go to zero at infinity. The velocity potential can be derived from (5.26). We notice that

$$\chi_e(z,t) = \frac{1}{\pi} \int_{-1}^{+1} \left[\left(\frac{\partial}{\partial t} + U \frac{\partial}{\partial \xi}\right) G\right] \log(\xi - z) \, d\xi, \tag{5.27}$$

from which follows that

$$\varphi_e(x,y,t) = \text{Re}\ \chi_e(z,t)$$

$$= \frac{1}{2\pi}\int_{-1}^{+1} \sigma(\xi,t)\ \log\left(\sqrt{(x-\xi)^2 + y^2}\right) d\xi. \tag{5.28}$$

Here

$$\sigma(\xi,t) = 2\left(\frac{\partial}{\partial t} + U\frac{\partial}{\partial \xi}\right) G(\xi,t),$$

is the source distribution.

Next, we determine a function $W_o(z,t)$ which is analytic in the complex z-plane except on $\mathcal{L}$ and $\mathcal{W}$ of the x-axis, where $\mathcal{W}$ is the downstream part of the axis (the wake). For $x < -1$, $y = 0$ we have $W_o^+(z,t) - W_o^-(z,t) = 0$ because the velocity is continuous. For $x \leqslant 1$, $y = 0$ we have

$$W_o^+ + W_o^- = -2i\left(\frac{\partial}{\partial t} + U\frac{\partial}{\partial x}\right) F,$$

while for $x > 1$

$$W_o^+ - W_o^- = [\varphi_{ox}] = u^+ - u^-,$$

which follows from (5.19).

To find $W_o(z,t)$, we use the theory of singular integral equations (see A.3), and represent $W_o(z,t)$ by the Cauchy integral:

$$W_o(z,t) = -\frac{1}{2\pi i}\int_{\mathcal{L}+\mathcal{W}} \frac{\gamma(\xi,t)}{\xi - z}\, d\xi, \tag{5.29}$$

where $\gamma(\xi,t)$ is a real function which is equivalent to a vortex distribution to be determined. By applying the Plemelj formulas to (5.29) we obtain:

$$W_o^+(\xi_0,t) - W_o^-(\xi_0,t) = -\gamma(\xi_0,t),$$

$$W_o^+(\xi_0,t) + W_o^-(\xi_0,t) = -\frac{1}{\pi i}\,⨍_{\mathcal{L}+\mathcal{W}} \frac{\gamma(\xi,t)}{\xi - \xi_0}\, d\xi, \tag{5.30}$$

where we use $⨍$ to denote the principal value of the integral.

From this it follows that

$$\gamma(\xi_0,t) = -[u](\xi_0,t), \qquad \text{for } \xi_0 > -1,$$

and for $\xi_0 > 1$, we know that (5.5) holds

$$\gamma(\xi_0,t) = \gamma(\xi_0 - Ut). \tag{5.31}$$

If we consider the situation at t = 0 the total circulation equals zero, we know that the total circulation remains zero. Hence we have

$$\int_{-1}^{\infty} \gamma(\xi,t)\, d\xi = \int_{\mathcal{L}} \gamma(\xi,t)\, d\xi + \int_{\mathcal{W}} \gamma(\xi,t)\, d\xi$$
$$= \Gamma(t) + \int_{1}^{\infty} \gamma(\xi,t)\, d\xi = 0, \tag{5.32}$$

where $\Gamma(t)$ is the circulation around the foil.
If we differentiate (5.32) with respect to t and use (5.31), we obtain

$$\frac{d\Gamma}{dt} = U\gamma(1,t). \tag{5.33}$$

If we start with a steady motion at t = 0, the vortex strength in the wake can be calculated easily by a step by step integration because of the relations (5.33) and (5.31). To assure that (5.30) has a unique solution for the unknown $\gamma(\xi,t)$, we furthermore require the Kutta condition which requires finite velocity at the trailing edge, which means that the pressure jump along the x-axis is a continuous function. Therefore we require

$$\gamma(1+0,t) = \gamma(1-0,t). \tag{5.34}$$

From (5.30) it follows that for $|x| < 1$ we have to solve a singular integral equation:

$$W_o^+ + W_o^- = -2iv(x,t) = -\frac{1}{\pi i}\int_{-1}^{+1} \frac{\gamma(\xi,t)}{\xi - x}\, d\xi - \frac{1}{\pi i}\int_{1}^{\infty} \frac{\gamma(\xi,t)}{\xi - x}\, d\xi, \tag{5.35}$$

where we consider $\gamma(\xi,t)$ to be known for $\xi > 1$. Hence the integral equation becomes

$$\lambda(x,t) = \frac{1}{i\pi} \int_{-1}^{+1} \frac{\gamma(\xi,t)}{\xi - x} \, d\xi. \tag{5.36}$$

Here $\lambda(x,t)$ is defined by:

$$-i\lambda(x,t) := 2v(x,t) + \frac{1}{\pi} \int_{1}^{\infty} \frac{\gamma(\xi - Ut)}{\xi - x} \, d\xi,$$

which can be computed stepwise.

By introducing

$$\Lambda(z,t) = \frac{1}{2\pi i} \int_{-1}^{+1} \frac{\gamma(\xi,t)}{\xi - z} \, d\xi, \tag{5.37}$$

the integral equation (5.36) can be solved by using the Plemelj formulas. We determine $\Lambda(z,t)$ such that

$$\Lambda^{+}(x,t) + \Lambda^{-}(x,t) = \lambda(x,t), \quad |x| < 1, \tag{5.38}$$

and Λ is analytic elsewhere. Clearly then

$$\gamma(x,t) = \Lambda^{+}(x,t) - \Lambda^{-}(x,t), \quad |x| < 1. \tag{5.39}$$

Observe that (5.38) is of the form (A.27) in the Appendix (A.3) with $G(s) = 1$.

For the special case when $\lambda = 0$ in (5.38), it can be easily found that

$$\Lambda_0(z,t) = \frac{1}{\sqrt{1 - z^2}},$$

is a solution of (5.38), where the branch cut is chosen between $-1 < x < 1$. From this particular solution Λ_0, it follows that the solution of (5.38) admits the representation (see (A.28)):

$$\Lambda(z,t) = \frac{\Lambda_0(z,t)}{2\pi i} \left\{\int_{-1}^{+1} \frac{\lambda(\xi,t)}{\xi - z} \Lambda_0^+(\xi,t)^{-1} \, d\xi + P(z,t)\right\} \tag{5.40}$$

where P(z,t) is an entire function of z. Consequently, by using the definition of Λ_0, the vortex strength becomes

$$\gamma(x,t) = \frac{1}{\pi i(1 - x^2)^{1/2}} \left\{\int_{-1}^{+1} \frac{\lambda(\xi,t)}{\xi - x} (1 - \xi^2)^{1/2} \, d\xi + P(x,t)\right\} \tag{5.41}$$

where

$$\lambda(\xi,t) = i \left\{2v(\xi,t) + \frac{1}{\pi} \int_1^\infty \frac{\gamma(s - Ut)}{s - \xi} \, ds\right\},$$

and

$$v(\xi,t) = \left(\frac{\partial}{\partial t} + U \frac{\partial}{\partial \xi}\right) F(\xi,t).$$

Because of the behaviour of the field at infinity, P(x,t) has to be a constant with respect to x, P(t). If we integrate (5.40) with respect to x along $\mathcal{L}$ we find

$$P(t) = \Gamma(t) = \int_{-1}^{+1} \gamma(x,t) \, dx. \tag{5.42}$$

If we start from a situation where the flow is symmetric like in the case of passing ships which we shall treat later on, the vortex distribution can be calculated. In that case $v_0(x,t) = 0$ in (5.35) for $t > 0$ and $v_0(x,t)$ is given for $t \geq 0$ and by means of a time step procedure $\gamma(x,t)$ can be determined uniquely from (5.5), (5.33), (5.34), (5.41) and (5.42). If the problem is not given as an initial value problem, then a more general procedure has to be followed.

We may now compute the lift force and moment acting on the airfoil. From Bernoulli's equation (5.7), we obtain the lift force as:

$$\begin{aligned} \mathcal{L} &= \rho \int_{-1}^{+1} \left\{\frac{\partial}{\partial t} [\varphi_o] + U \frac{\partial}{\partial x} [\varphi_o]\right\} dx \\ &= -\rho \int_{-1}^{+1} \left\{\frac{\partial}{\partial t} \int^x \gamma ds + U\gamma\right\} dx, \end{aligned} \tag{5.43}$$

and the moment about x = 0 becomes

$$
\begin{aligned}
M &= \rho \int_{-1}^{+1} \left\{ \frac{\partial}{\partial t} [\varphi_o] + U \frac{\partial}{\partial x} [\varphi_o] \right\} x dx \\
&= -\rho \int_{-1}^{+1} \left\{ \frac{\partial}{\partial t} \int^{x} \gamma ds + U\gamma \right\} x dx. \qquad (5.44)
\end{aligned}
$$

Once $\gamma(x,t)$ is known along the foil, L and M may be computed easily.

5.2 Slender Body Theory

We now consider the situation where a slender body is placed in between two parallel walls with a distance a (see Figure 5.2). The uniform flow is given by $\underset{\sim}{U} = (0,V,0)$ as $|\underset{\sim}{x}| \to \infty$ with constant V. In particular, we assume that $a = O(\varepsilon)$, where ε is the slenderness parameter $\varepsilon = B/L$, L the length and B the width of the slender body. The region where both y and z are small will be called the inner region, while the region where $y = O(1)$ is called the outer region. As in (5.1), we may stretch the coordinates y and z to formulate the inner problem, which will become a two-dimensional problem. This leads to the following boundary value problem for the first order approximation of the potential φ^i in the inner region:

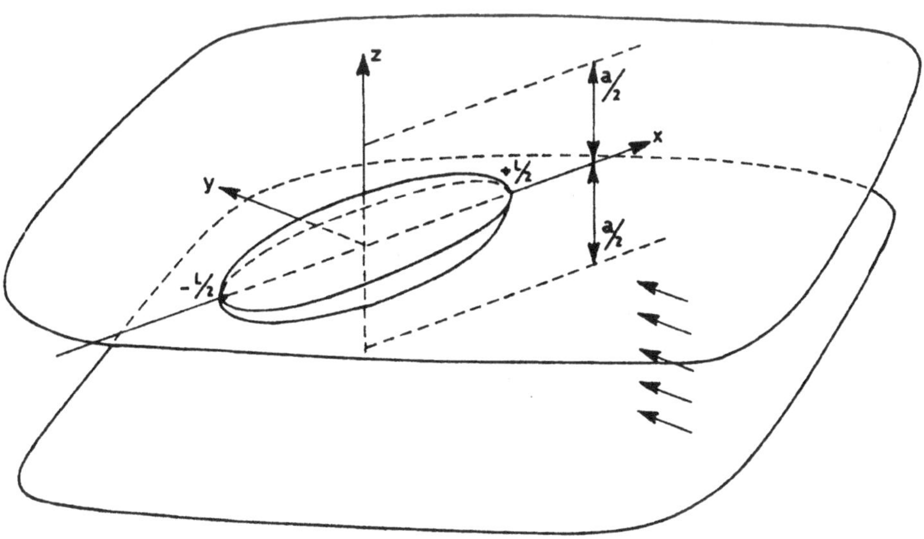

Figure 5.2

$$\varphi^i_{yy} + \varphi^i_{zz} = 0, \qquad -\frac{a}{2} \leq z \leq \frac{a}{2}, \quad -\infty < y < \infty,$$

$$\frac{\partial \varphi^i}{\partial n} = 0, \quad \text{on the boundary of the slender body } S,$$

$$\varphi^i \to V_0(x)(y \mp C(x)), \qquad \text{as } y \to \pm\infty, \tag{5.45}$$

where $V_0(x)$ will be determined from the matching condition later. Here $C(x)$ is usually referred to as the 'blockage constant'.

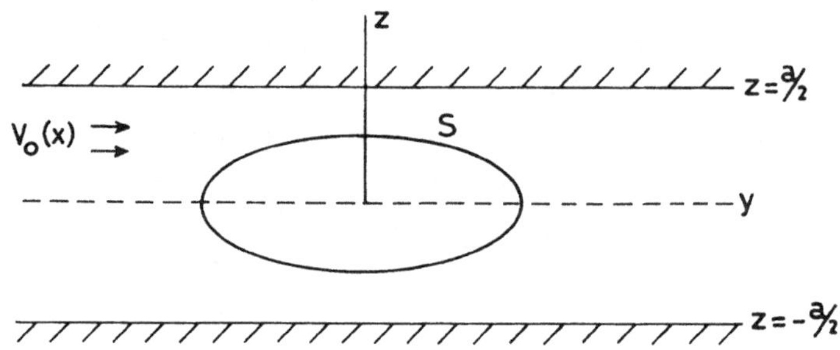

Figure 5.3

To illustrate the significance of $C(x)$, we consider the special case when a cascade of dipoles with orientation in the y-direction is super-imposed on the current V_0, and these dipoles are situated at $y = 0$, $z = na$ with integer n. The strength of the dipoles is chosen in such a way that the flow due to one dipole in a flow V_0 describes the flow around a circular cylinder with radius ρ_1. Hence the cascade flow is the flow between two parallel walls at $z = \pm a/2$ around a cylinder whose shape differs slightly from a circle.

The potential φ^i becomes

$$\varphi^i(y,z) = V_0(x)\left\{y + \sum_{n=-\infty}^{\infty} \frac{\rho_1^2 y}{y^2 + (z - na)^2}\right\}.$$

For large values of $|y|$ and $|z| < a/2$, we see that

$$\varphi^{i}(y,z) \approx V_0(x)\left(y \pm \frac{S}{a}\right), \tag{5.46}$$

where $S = \pi\rho_1^2$ the surface area, and the blockage constant $C = S/a$. In general, a relation between the surface area S, the virtual mass m_{yy} and the blockage coefficient C can be derived such that

$$m_{yy} = -\rho S + 2\rho aC. \tag{5.47}$$

In particular, for a flat plate of length ℓ, perpendicular to the flow between two parallel walls, m_{yy} becomes

$$m_{yy} = -2\rho\, a^2/\pi \log \cos \frac{\pi\ell}{2a} = 2\rho aC,$$

since S = 0 in (5.47). Hence, for most shapes the blockage coefficient may be calculated and we consider it known.

The potential in the outer region is obtained by a formal stretching of the z-coordinate. It turns out that up to the first and second approximation, φ^0 has to be a function of x and y only, and is a solution of the two-dimensional Laplace equation:

$$\varphi^0_{xx} + \varphi^0_{yy} = 0.$$

The boundary conditions are

$$\varphi^0 \sim Vy, \qquad \text{as } y \to \infty,$$

and

$$\frac{\partial\varphi^0}{\partial y} = V_0(x), \qquad \text{for } -1 < x < 1 \text{ and } y = 0.$$

By introducing the complex velocity function $f(\zeta) = u - iv$ where $\zeta = x + iy$, these conditions can be simplified as

$$f(x \pm i0) = -iV_0(x), \qquad \text{for } |x| < 1,$$

$$f(\zeta) = -iV, \qquad |\zeta| \to \infty,$$

and $f(\zeta)$ is analytic in the complex ζ-plane such that

$$f^+(x) + f^-(x) = -2iV_0(x), \qquad \text{for } |x| < 1. \qquad (5.48)$$

The flow around the endpoints may be singular, which leads to the choice of the solution of (5.48) for $V_0 = 0$,

$$f_0(\zeta) = (1 - \zeta^2)^{-1/2}, \qquad (5.49)$$

and the solution of (5.48)

$$f(\zeta) = -\frac{f_0(\zeta)}{\pi} \int_{-1}^{+1} \frac{V_0(\xi) f_0^+(\xi)^{-1}}{\xi - \zeta} d\xi + f_0(\zeta)P(\zeta), \qquad (5.50)$$

where, to fulfill the condition at infinity, $P(\zeta)$ has the form:

$$P(\zeta) = V\zeta + P(0).$$

We now match the velocity in the x-direction of the outer and inner potentials and obtain from (5.49):

$$u(x,\pm 0) = \mp \frac{1}{\pi(1 - x^2)^{1/2}} \int_{-1}^{+1} \frac{V_0(\xi)(1 - \xi^2)^{1/2}}{\xi - x} d\xi$$

$$\mp \frac{Vx + P(0)}{(1 - x^2)^{1/2}} = \pm \frac{d}{dx} (V_0 C), \qquad (5.51)$$

where $P(0)$ and $V_0(x)$ remain to be determined. To do so, we integrate the integral equation (5.51) with respect to x from -1 to x:

$$V_0(x)C(x) - V_0(-1)C(-1) = -\frac{1}{\pi} \int_{-1}^{+1} V_0(\xi)\, K(x,\xi)\, d\xi$$

$$+ V(1 - x^2)^{1/2} - P(0)\, (\tfrac{1}{2}\pi + \arcsin x), \qquad (5.52)$$

where $K(x,\xi)$ is a symmetric kernel of the form:

$$K(x,\xi) = \int_{-1}^{x} \frac{(1 - \xi^2)^{1/2}\, dx}{(1 - x^2)^{1/2}(\xi - x)} \qquad (5.53)$$

$$= \frac{1}{2} \log \left[\frac{1 - \xi x + (1 - \xi^2)^{1/2}(1 - x^2)^{1/2}}{1 - \xi x - (1 - \xi^2)^{1/2}(1 - x^2)^{1/2}}\right].$$

For slender body with rounded endpoints where $C(\pm 1) = 0$ it follows that $P(0) = 0$. If the cross flow problem is combined with a lateral flow, a Kutta condition may be imposed at $x = 1$. In this case, it can be shown that

$$P(0) = -V \pm 1/\pi \int_{-1}^{+1} V_0(\xi) \left(\frac{1+\xi}{1-\xi}\right)^{1/2} d\xi. \tag{5.54}$$

With $P(0)$ determined, (5.52) is a Fredholm integral equation of the second kind for the unknown $V_0(x)$. For large values of $C(x)$, we can find an approximate solution of (5.52) in the form:

$$V_0(x)C(x) = V(1 - x^2)^{1/2}, \tag{5.55}$$

in the case $P(0) = 0$, while

$$V_0(x)C(x) = V\left[(1 - x^2)^{1/2} + \frac{1}{2}\pi + \text{arc sin } x\right] \tag{5.56}$$

in the latter case, where the lift force and moment may be calculated. We find

$$L = 2\rho a\, V_0(1)C(1),$$

and

$$M = 2\rho a\, V_0(1)C(1) - \rho V \forall - 2\rho a \int_{-1}^{+1} V_0(x)C(x)\, dx,$$

where $\forall = \int_{-1}^{+1} S\, dx$ is the volume of the body.

5.3 Free Surface Effects

In the preceding sections, we dealt with flow problems in an infinite fluid medium. In this section, we derive equations for the shallow water ship problem where a free surface plays an important role. We consider the steady flow around a slender ship and assume that the angle of attack of the undisturbed flow is small. This situation occurs quite often near shallow water harbour entrances or a ship manoeuvring at shallow water where the variations in the course of the ship are with such a time scale that the flow may be considered to be stationary for each time interval. To be more specific, we choose a coordinate system fixed to the ship, the positive x-axis is directed in the direction of the projection of the undisturbed incoming flow and the incoming flow makes an angle α with the x-axis. In Figure 5.4, B^* denotes the width of the ship, T^* the draft of the ship, and L the length of the ship. The depth of the water is denoted by h^*.

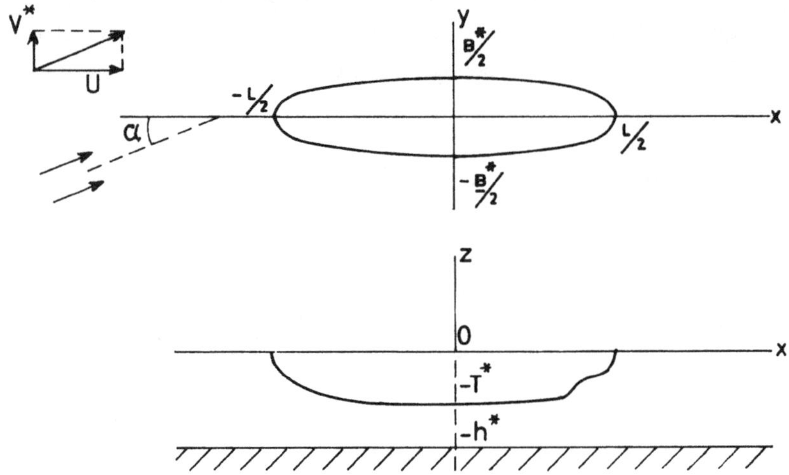

Figure 5.4

We assume that the velocity of the uniform flow is given by $\underset{\sim}{U} = (U, V^*, 0)$ where U and V^* are constants. The equation for the ship hull is given by

$$y = \pm f(x,z) = \pm \varepsilon F(x,z),$$

where ε is defined by $\varepsilon = B^*/L$. The other dimensions are chosen such that h^*/L, T^*/L and $V^*/U = O(\varepsilon)$, as $\varepsilon \to 0$. Furthermore, the Froude number defined with the depth as the characteristic length is denoted by

$$F_{h^*}^2 = |\underset{\sim}{U}|^2/gh^* = O(1), \qquad \text{as } \varepsilon \to 0,$$

and the Froude number with respect to the length L is given by

$$F_L = |\underset{\sim}{U}|/\sqrt{gL} = O(\varepsilon^{1/2}).$$

First, we consider the outer region at a finite distance from the ship where

$$x, y = O(1) \quad \text{and} \quad z = O(\varepsilon).$$

Denoting the unperturbed potential by φ_0^0, $\varphi_0^0 = Ux + V^*y$, and the perturbed potential by φ^0, we have

$$\underset{\sim}{u} = \nabla\left(\varphi_0^0 + \varphi^0\right) =: \nabla\Phi^0, \tag{5.57}$$

$$\Delta\varphi^0 = 0.$$

At the flat bottom $z = -h^*$, we have

$$\frac{\partial\varphi^0}{\partial z} = 0. \tag{5.58}$$

At the free surface $z = \zeta(x,y)$, we have the kinematic condition:

$$\varphi_z^0 - \varphi_x^0\zeta_x - \varphi_y^0\zeta_y = U\zeta_x + V^*\zeta_y \tag{5.59}$$

and dynamic condition:

$$-g\zeta = U\varphi_x^0 + V^*\varphi_y^0 + \frac{1}{2}\left({\varphi_x^0}^2 + {\varphi_y^0}^2 + {\varphi_z^0}^2\right). \tag{5.60}$$

Introducing the stretching variable $Z := z/\varepsilon$, we seek $\varphi^0(x,y,Z,\varepsilon)$ as a regular perturbation series in the form:

$$\varphi^0(x,y,Z,\varepsilon) = \varepsilon\varphi_1(x,y,Z) + \varepsilon^2\varphi_2(x,y,Z) + \ldots..$$

From the Laplace equation which becomes now in terms of Z,

$$\varphi_{ZZ}^0 = -\varepsilon^2(\varphi_{xx}^0 + \varphi_{yy}^0) =: -\varepsilon^2\Delta_2\varphi^0,$$

we find for φ_i such that

$$\varphi_{iZZ} = \begin{cases} 0, & i = 1,2 \\ -\Delta_2\varphi_{i-2}, & i = 3,4.\dots, \end{cases}$$

from which it follows that

$$\begin{aligned} \varphi_1 &= \psi_1(x,y) \\ \varphi_2 &= \psi_2(x,y) \\ \varphi_3 &= \psi_3(x,y) - \tfrac{1}{2}(Z+h)^2\,\Delta_2\psi_1(x,y), \end{aligned} \tag{5.61}$$

with $h = h^*/\varepsilon$. The ψ_i can be determined in the following way.

We begin with the dynamic free surface condition (5.60). This leads to the relation:

$$-g\zeta = \varepsilon U\psi_{1x} + \varepsilon^2\left(U\psi_{2x} + V\psi_{1y} + \psi_{1x}^2 + \psi_{1y}^2\right) + O(\varepsilon^3), \tag{5.62}$$

with $V = V^*/\varepsilon$. Then equation (5.62) together with the fact that $F_{h^*}^2 = O(1)$ implies that one may consider ζ in the form:

$$\zeta = \varepsilon^2\zeta_2 + \varepsilon^3\zeta_3 + \dots\dots,$$

and obtain from (5.62) by equating like powers of ε:

$$\begin{aligned} \zeta_2 &= -\frac{U}{\varepsilon g}\psi_{1x} = -\frac{\ell}{U}\psi_{1x}, \\ \zeta_3 &= -\frac{\ell}{U^2}\left\{U\psi_{2x} + V\psi_{1y} + \frac{1}{2}\left(\psi_{1x}^2 + \psi_{1y}^2\right)\right\}, \end{aligned} \tag{5.63}$$

where ℓ is of $O(1)$ and defined by $U^2/g = \varepsilon\ell$. Now it follows from the kinematic condition (5.59) together with (5.61) and (5.63) that

$$\begin{aligned} -\varepsilon^2 h\Delta_2\psi_1 - \varepsilon^3(\zeta_2\Delta_2\psi_1 + h\Delta_2\psi_2) + O(\varepsilon^4) \\ = \varepsilon^3 U\zeta_{2x} + \varepsilon^3(U\zeta_{3x} + V\zeta_{2y} + \zeta_{2x}\psi_{1x} + \zeta_{2y}\psi_{1y}) \\ + O(\varepsilon^4). \end{aligned}$$

Consequently, we obtain:

$$\Delta_2\psi_1 = -\frac{U}{h}\zeta_{2x},$$
$$\Delta_2\psi_2 = -\frac{\zeta_2}{h}\Delta_2\psi_1 - \frac{1}{h}\left\{U\zeta_{3x} + V\zeta_{2y} + \zeta_{2x}\psi_{1x} + \zeta_{2y}\psi_{1y}\right\}. \tag{5.64}$$

Eliminating ζ_2 and ζ_3 from (5.64) and (5.63) yields:

$$\left(1 - F_{h*}^2\right)\psi_{1xx} + \psi_{1yy} = 0, \text{ and}$$
$$\left(1 - F_{h*}^2\right)\psi_{2xx} + \psi_{2yy} = \Omega_2(x,y), \tag{5.65}$$

where $\Omega_2(x,y)$ is a function of ζ_2 and ψ_1 only. Equation (5.65) is the linearized shallow water equation which is a partial differential equation of the mixed type depending on the value of F_{h*}. We confine only to the elliptic case where $F_{h*} < 1$. Boundary conditions for ψ_1 follow from a matching procedure at $y = 0$ with the inner solution.

We now turn to the solution of (5.65) and write

$$\psi_1(x,y) = \psi_1^e(x,y) + \psi_1^o(x,y),$$

where ψ_1^e is an even function of y while ψ_1^o is odd in y. The even part may be written as a source distribution

$$\psi_1^e(x,y) = \frac{1}{2\pi\sqrt{1 - F_{h*}^2}}\int_{-\infty}^{\infty}\sigma(\xi) \times \log\left\{(x-\xi)^2 + \left(1 - F_{h*}^2\right)y^2\right\}^{1/2} d\xi \tag{5.66}$$

and the odd part may be written as a vortex distribution

$$\psi_1^o(x,y) = \frac{1}{2\pi\sqrt{1 - F_{h*}^2}}\int_{-\infty}^{\infty}\gamma(\xi)\cdot\arctan\left(\frac{y\sqrt{1 - F_{h*}^2}}{x-\xi}\right) d\xi. \tag{5.67}$$

Because of the proposed matching at $y = 0$, we must make a series expansion for small values of y for both ψ_1^e and ψ_1^o. We find the Taylor series expansion:

$$\psi_1^e(x,y) \approx \psi_1^e(x,0) \; |y| \; \frac{\sigma(x)}{2}$$

$$- \frac{1}{2} y^2 \left(1 - F_{h*}^2\right) \psi_{1xx}^e(x,0) + \ldots, \tag{5.68}$$

where

$$\psi_1^e(x,0) = \frac{1}{2\pi\sqrt{1 - F_{h*}^2}} \int_{-\infty}^{\infty} \sigma(\xi) \log |x - \xi| \, d\xi.$$

For the odd part, we find

$$\psi_1^o(x,y) \approx + \operatorname{sgn} y \; \Gamma(x)$$

$$+ \frac{y}{2\pi} \int_{-\infty}^{\infty} \frac{\gamma(\xi)}{x - \xi} \, d\xi + O(y^2), \tag{5.69}$$

where $\Gamma(x) = - \frac{1}{2} \int_{-\infty}^{x} \gamma(\xi) \, d\xi = - \frac{1}{2} \int_{-\infty}^{\infty} \gamma(\xi) \, H(x - \xi) \, d\xi$ with $H(x - \xi)$ denoting the Heaviside step function.

Now we are in a position to consider the expansion of the total outer field $\Phi^0 = \varphi_0^0 + \varphi^0$ for small values of y, where φ_0^0 and φ^0 are unperturbed and perturbed potentials as in (5.57). Collecting the results (5.57), (5.61), (5.68) and (5.69), we finally arrive at the expression:

$$\Phi^0 \approx Ux + \varepsilon\psi_1^e(x,0) \; \pm \; \varepsilon\Gamma(x)$$

$$+ y \left(V^* + \frac{\varepsilon}{2\pi} \int_{-\infty}^{\infty} \frac{\gamma(\xi) \, d\xi}{x - \xi}\right) \pm y \, \frac{\varepsilon\sigma(x)}{2} + O(y^2)$$

$$=: Ux + \varphi_1^e(x) \; \pm \; [\varphi_1^o](x)$$

$$+ yV^o(x) \; \pm \; y[V^e](x) + O(y^2), \qquad \text{for } y \gtrless 0, \tag{5.70}$$

with φ_1^e, $[\varphi_1^o]$, V^o and $[V^e]$ defined accordingly. It should be emphasized that the densities σ and γ in the above formulation remain to be determined. For this purpose, we have to consider the problem in the inner region.

In the inner region we have

$$x = O(1), \quad y = O(\varepsilon) \quad \text{and} \quad z = O(\varepsilon).$$

Hence we introduce the stretched variables $Y = y/\varepsilon$ and $Z = z/\varepsilon$, in terms of which the Laplace equations now become

$$\varepsilon^2\Phi^i_{xx} + \Phi^i_{YY} + \Phi^i_{ZZ} = 0,$$

where we have adapted the notion Φ^i for the inner potential $\Phi^i = \Phi^i(x,Y,Z,\varepsilon)$. We assume that Φ^i admits the regular expansion in the form:

$$\Phi^i(x,Y,Z,\varepsilon) = \Phi_0(x,Y,Z) + \varepsilon\Phi_1(x,Y,Z) + O(\varepsilon^2), \tag{5.71}$$

and substituting it into the Laplace equation, we obtain:

$$\tilde{\Delta}_2\Phi_i = \begin{cases} 0, & i = 1,2 \\ -\Phi_{i-2}, & i = 3,4\ldots, \end{cases}$$

where $\tilde{\Delta}_2 := (\partial^2/\partial Y^2) + (\partial^2/\partial Z^2)$. The free surface conditions (5.59) and (5.60) yield for the first two approximations:

$$\Phi_{0Z} = 0, \quad \Phi_{1Z} = 0, \quad \text{at } Z = 0.$$

Again we split up the inner potential Φ^i into Φ^{ie} and Φ^{io}, the even and the odd part of Φ^i. The even potential relates to the thickness problem; since the even part of the unperturbed flow in the outer region equals Ux, we take $\Phi^e_0 = Ux$. Furthermore, Φ^{ie} has to fulfill the boundary condition on the ship hull, i.e.,

$$\frac{\partial\Phi^{ie}}{\partial n} = 0, \quad \text{on } y = \pm f(x,z) = \pm\varepsilon F(x,Z). \tag{5.72}$$

We now define $\underset{\sim}{N}^*$ to be the projection of the normal $\underset{\sim}{n}$ on the plane x = constant (see Figure 5.5). From (5.71) for $y > 0$,

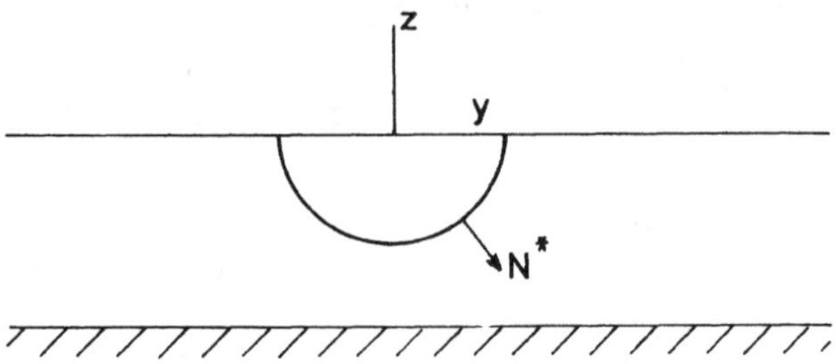

Figure 5.5

we have

$$\Phi_x^e f_x + \Phi_z^e f_z - \Phi_y^e = 0.$$

We also have

$$\frac{\partial \Phi^{ie}}{\partial N^*} = \frac{\Phi_y^{ie} - \Phi_z^{ie} f_z}{\sqrt{1 + f_z^2}},$$

from which it follows that

$$\frac{\partial \Phi^{ie}}{\partial N^*} = \frac{\Phi_x^{ie} f_x}{\sqrt{1 + f_z^2}}. \qquad (5.73)$$

Inserting (5.71) into (5.73) and defining the normal $\underset{\sim}{N}$ in terms of the stretched coordinates Y,Z, $\frac{\partial}{\partial N^*} = \frac{1}{\varepsilon} \frac{\partial}{\partial N}$, we then find with $f = \varepsilon F$ that

$$\frac{\partial \Phi_1^e}{\partial N} = 0,$$

$$\frac{\partial \Phi_2^e}{\partial N} = \frac{U F_x}{\sqrt{1 + F_z^2}} =: V_N.$$

Therefore, Φ_1^e and Φ_2^e take the forms:

$$\Phi_1^e = g_1(x), \qquad (5.74)$$

$$\Phi_2^e = g_2(x) + \Phi_2^*(x,Y,Z).$$

This inner solution has to be matched with the outer one. Therefore Φ_2^* must be known for large values of Y and there is no need to solve the equation for Φ_2^*. For this purpose, we assume Φ_2^* has the asymptotic behaviour:

$$\Phi_2^* \sim V|Y|, \qquad \text{as } Y \to \infty,$$

where V may be calculated simply by conservation of mass. We proceed as follows.

As indicated in Figure 5.6, the apparent flux out of the ship is given by

$$\int V_N dl = US'(x),$$

where S' is the derivative of the surface area of the stretched cross section. Hence

$$V = \frac{U}{2h} S'(x), \qquad \text{for } |x| < L/2.$$

We now match the even part of the inner and outer solution (5.68), (5.74), and we find

$$\sigma(x) = \frac{U}{h} S'(x) \tag{5.75}$$

and

$$g_1(x) = \psi_1^e(x,0)$$

$$= \frac{U}{2\pi h\sqrt{1 - F_{h*}^2}} \int_{-L/2}^{L/2} S'(\xi) \log |x - \xi| \, d\xi.$$

The odd part of the solution relates to the cross flow $V^o(x)y$ given by (5.70). The inner problem we have to consider is the same as in Section 5.2, if we reflect the physical domain with respect to y = 0 (see Figure 5.7).

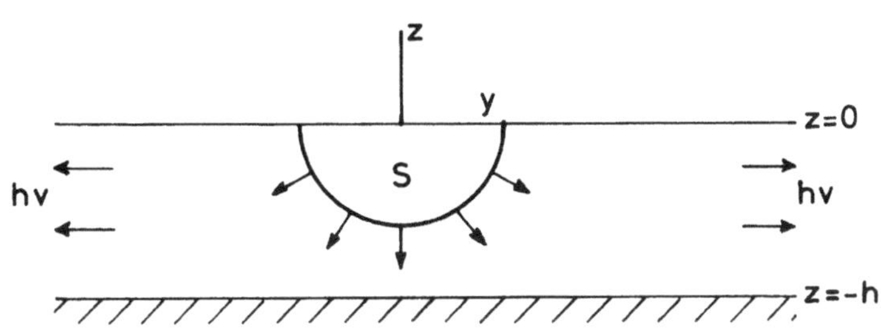

Figure 5.6

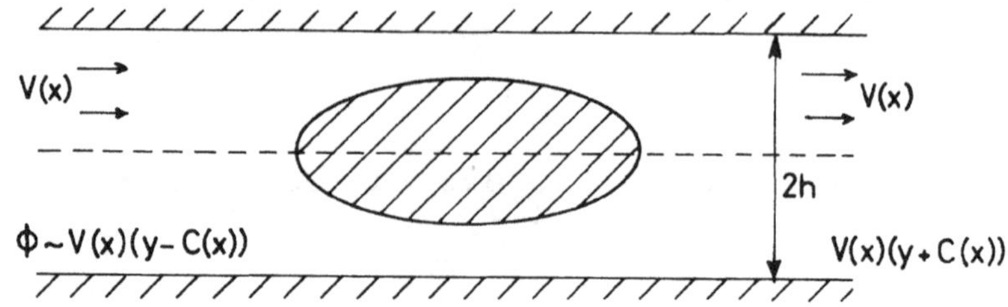

Figure 5.7

In this figure, C(x) is the known blockage function as introduced in (5.45) and V(x) is to be determined by the matching procedure below. The far field behaviour of the inner solution can be matched with the local behaviour of the outer solution. We obtain from (5.70):

$$[\varphi_1^o](x) = V(x)\ C(x), \text{ and}$$
$$v^o(x) = V(x).$$

Consequently, we have

$$-\frac{\varepsilon}{2}\int_{-\infty}^{\infty} \gamma(\xi)\ H(x-\xi)\ d\xi = V(x)\ C(x) \qquad (5.77)$$

and

$$V^* + \frac{\varepsilon}{2\pi}\int_{-\infty}^{\infty} \frac{\gamma(\xi)}{x-\xi}\ d\xi = V(x), \qquad (5.78)$$

from which it follows that $\gamma(\xi)$ has to be a solution of the following singular integral equation:

$$\frac{\varepsilon}{2\pi}\int_{-\infty}^{\infty} \left\{\frac{\gamma(\xi)}{x-\xi} + \frac{\gamma(\xi)H(x-\xi)}{C(x)}\right\} d\xi$$
$$= -V^* := -\varepsilon V. \qquad (5.79)$$

We can easily obtain an integral equation for V(x), because differentiation of (5.77) with respect to x leads to

$$\varepsilon\gamma(x) = -2(VC)'.$$

Since in this section we consider the case of the steady motion, it follows that $\gamma(\xi) = 0$, for $|x| > L/2$.

We shall show that (5.79) can be transposed into a Fredholm integral equation of the second kind, which can be solved numerically.

Consider:

$$\frac{1}{\pi i} \int_{-L/2}^{L/2} \frac{\gamma(\xi)}{x - \xi} d\xi = 2iV + \frac{i}{C(x)} \int_{-L/2}^{x} \gamma(\xi) \, d\xi := iF_{\gamma}(x). \tag{5.80}$$

We now use the theory of Hilbert transforms (see A.3) again and consider

$$\frac{1}{\pi i} \int_{-L/2}^{L/2} \frac{\gamma(\xi)}{x - \xi} d\xi = iF_{\gamma}(x).$$

We apply the Kutta condition at x = L/2 to obtain:

$$\gamma(x) = \frac{1}{\pi} \sqrt{\frac{(x - L/2)}{(x + L/2)}} \int_{-L/2}^{L/2} \sqrt{\frac{(\zeta + L/2)}{(\zeta - L/2)}} \frac{F_{\gamma}(\zeta)}{x - \zeta} d\zeta. \tag{5.81}$$

Equation (5.81) now is a Fredholm integral equation of the second kind for $\gamma(x)$. It can be shown that the kernel is a square integrable and hence has a unique solution. This unique solution can be found numerically. In particular, if $C(x) \gg 1$, one can define a sequence $\{\gamma_i\}$ by the iteration procedure.

$$\gamma_i(x) = \frac{1}{\pi} \sqrt{\frac{(x - L/2)}{(x + L/2)}} \int_{-L/2}^{L/2} \sqrt{\frac{(\zeta + L/2)}{(\zeta - L/2)}} \frac{1}{x - \zeta} \times \left[2V + \frac{1}{C(\zeta)} \int_{-L/2}^{\zeta} \gamma_{i-1}(\xi) \, d\xi \right] d\zeta. \tag{5.82}$$

This sequence can be used to approximate the exact solution.

In the case of total blockage $C(x) = \infty$ we have obtained the same result as in the 2-D airfoil theory for steady flow. Equation (5.82) with $C = \infty$ is identical to (5.41) if $P(x)$ is determined so that the Kutta condition $\gamma(L/2) = 0$ is fulfilled. It will be clear that the theory of this section may be extended to non-steady flow problems in a similar way as in Section 5.1. We then have to use the same conditions to obtain a unique vortex distribution.

To obtain the heave force and pitch moment we shall integrate the local pressure distribution along the hull. The local field up to first order in ε is described by

$$\Phi^i \approx Ux + \varepsilon\psi_1^e(x,0) \pm \varepsilon\Gamma(x), \tag{5.83}$$

and for the determination of the heave and pitch, we may disregard the odd term $\pm\varepsilon\Gamma(x)$. We use Bernoulli's equation and find that the even part of the first order pressure along the ship is given by

$$p^e \sim -\rho U\varepsilon\psi_{1x}^e(x,0). \tag{5.84}$$

The heave force becomes

$$F_3 = -\rho U\varepsilon \int_{-L/2}^{L/2} \psi_{1x}^e B(x)\, dx, \tag{5.85}$$

and the pitch moment

$$F_5 = \rho U\varepsilon \int_{-L/2}^{L/2} \psi_{1x}^e B(x)\, x\, dx, \tag{5.86}$$

where $B(x)$ is the beam of the ship at the location x.

The sway force and yaw moment can be calculated similarly. From (5.83), contribution from the odd part $\pm\varepsilon\Gamma(x)$ leads to:

$$p^0(x) \sim \pm\rho U\varepsilon\Gamma_x(x).$$

Hence a sway force (lift) is given by

$$F_2 = -\rho Uh\varepsilon \int_{-L/2}^{L/2} \gamma(\xi)\, d\xi. \tag{5.87}$$

The yaw moment now becomes

$$F_6 = -\rho U h \varepsilon \int_{-L/2}^{L/2} \xi \gamma(\xi)\, d\xi. \tag{5.88}$$

The sway force and yaw moment may also be calculated by means of the Blasius formulas. In that case, we write the outer potential as the real part of the complex potential

$$\chi(z) = \Phi(z) + i\psi(z), \text{ where } z = x + iy.$$

The forces acting on the ship are then

$$F_1 - iF_2 = \frac{i\rho}{2} \oint_C \left(\frac{dW}{dz}\right)^2 dz, \tag{5.89}$$

and the moment

$$F_6 = \text{Re} \left\{- \frac{1}{2} \rho \oint_C z \left(\frac{dW}{dz}\right)^2 dz\right\}, \tag{5.90}$$

where the contour C is a closed contour around the object. In our case we considered

$$W(z) = (U + iV^*)z + \text{source dist.} + \text{vortex dist.},$$

and if we equate the contribution of the residues, it follows that $F_1 = 0$ while F_2 and F_6 have the forms as given by (5.87) and (5.88), respectively. However, we would like to mention that this derivation only holds for steady flow fields.

5.4 Ships in a Channel

In this section we will show how we can apply the theory developed in the previous section to the situation of a ship in a channel. Since most of the analysis we consider here is exactly the same as before, we omit the detail. We consider the case that the side walls of the channel are in the outer region. The outer field may be considered as a superposition of sources and vortices. First of all we shall give a derivation of the potential due to a source of unit strength situated between two parallel walls.

We introduce the complex variable

$$z = x + i\beta y, \quad \text{with } \beta = \sqrt{1 - F_h^2},$$

where F_h is the Froude number with respect to the depth h of the channel. Throughout the section, the side walls of the channel are at y = a and y = -b. The total width of the channels is denoted by w as shown in Figure 5.8.

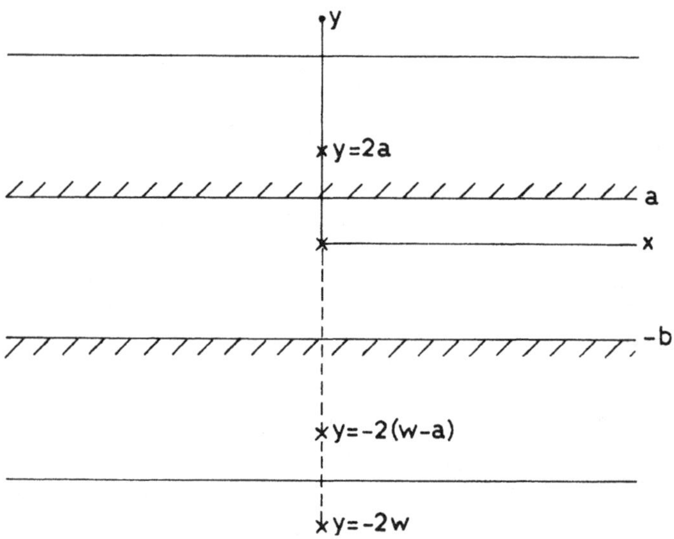

Figure 5.8

We consider a source with unit strength at the origin. The field due to this source may be obtained by means of reflection with respect to the walls. In this case, the complex potential denoted by χ(z) has the form:

$$\chi(z) = \varphi + i\psi = \frac{1}{2\pi}\log z + \frac{1}{2\pi}\sum_{n=1}^{\infty}\log(z - 2\beta inw)$$

$$+ \frac{1}{2\pi}\sum_{n=1}^{\infty}\log(z + 2\beta inw) + \frac{1}{2\pi}\log(z - 2\beta ia)$$

$$+ \frac{1}{2\pi}\sum_{n=1}^{\infty}\log(z - 2i\beta(nw + a))$$

$$+ \frac{1}{2\pi}\sum_{n=1}^{\infty}\log(z + 2i\beta(nw - a)). \qquad (5.91)$$

We now consider the first three terms in (5.91) and write

$$\begin{aligned}\chi_1 &= \frac{1}{2\pi}\log z + \frac{1}{2\pi}\sum_{n=1}^{\infty}\log(z - 2i\beta nw)\\ &\quad + \frac{1}{2\pi}\sum_{n=1}^{\infty}\log(z + 2i\beta nw)\\ &= \frac{1}{2\pi}\log\left[z\prod_{n=1}^{\infty}\left(1 - \frac{z^2}{(2i\beta nw)^2}\right)\cdot\text{constant}\right].\end{aligned}$$

The function $z\prod_{n=1}^{\infty}\left(1 - \frac{z^2}{(2i\beta nw)^2}\right)$ is an entire function with zero's at $z = \pm 2inw$. The lemma of Weierstrass allows us to write

$$\begin{aligned}z\prod_{n=1}^{\infty}\left(1 - \frac{z^2}{(2i\beta nw)^2}\right) &= \frac{2i\beta w}{\pi}\sin\frac{z\pi}{2i\beta w}\\ &= \frac{2\beta w}{\pi}\sinh\frac{z\pi}{2\beta w}.\end{aligned}$$

Hence up to an additive constant we can simplify χ_1 so that

$$\chi_1 = \frac{1}{2\pi}\log[\sinh k_0 z],$$

with $k_0 = \frac{\pi}{2\beta w}$, $z = x + i\beta y$ and $\beta = \sqrt{1 - F_h^2}$. The remaining three terms in (5.91) lead to a similar result, and the complex potential due to a unit source at $z = \zeta$ becomes

$$\begin{aligned}G_s(z;\zeta) &= \frac{1}{2\pi}\Big\{\log\Big[\sinh k_0(z - \zeta)\Big]\\ &\quad + \log\Big[\sinh(k_0(z - \zeta) - i\alpha)\Big]\Big\},\end{aligned} \tag{5.92}$$

with $\alpha = \frac{\pi a}{w}$. The complex potential due to a unit vortex at the point $z = \zeta$ becomes

$$\begin{aligned}G_v(z;\zeta) &= -\frac{i}{2\pi}\Big\{\log\Big[\sinh k_0(z - \zeta)\Big]\\ &\quad - \log\Big[\sinh(k_0(z - \zeta) - i\alpha)\Big]\Big\}.\end{aligned} \tag{5.93}$$

The perturbation potential in the outer region (see (5.57)) can now be written as a distribution of sources and vortices along the axis of the ship:

$$\varphi^0(x,y) = \mathrm{Re}\int_{-L/2}^{L/2}\left[\sigma(\xi)G_s(z;\xi) + \gamma(\xi)G_v(z;\xi)\right]d\xi, \tag{5.94}$$

where ξ is the coordinate along the axis of the ship. Here again σ and γ are the unknown densities as in (5.66) and (5.67). They will be determined by the matching conditions from the inner potential Φ^i as in Section 5.3. To do so, we introduce, for convenience, the (s,t) coordinates fixed to the ship and let (U,V) be the components of the uniform flow U with respect to the coordinates (s,t) (as shown in Figure 5.9). Furthermore we assume the angle of attack is small and expand (5.94) for small t.

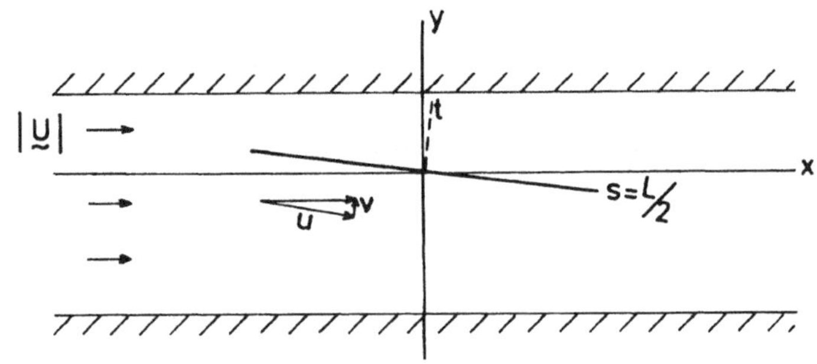

Figure 5.9

This leads after some lengthy calculations to

$$\varphi^i(s,t) \sim Us + Vt + \varphi_1(s) \pm [\varphi_1](s)$$
$$+ t\Big(V(s) \pm [V](s)\Big), \tag{5.95}$$

where

$$\varphi_1(x) = \frac{1}{2\pi}\int_{-L/2}^{L/2} d\xi\,\Bigg(\sigma(\xi)\left\{\log\left[\sinh k_0(s-\xi)\right]\right.$$
$$\left.+ \frac{1}{2}\log\left[\sinh k_0(s-\xi) + \sin^2\alpha\right]\right\}$$
$$+ \gamma(\xi)\arctan\left[\frac{\cosh\,[k_0(s-\xi)]\,\sin\alpha}{\sinh\,[k_0(s-\xi)]\,\cos\alpha}\right]\Bigg);$$

$$[\varphi_1] = -\frac{1}{2}\int_{-L/2}^{L/2} \gamma(\xi)H(s-\xi)\,d\xi;$$

$$V(s) = \frac{1}{4w}\int_{-L/2}^{L/2} d\xi \left\{\left(\frac{-\sigma(\xi)\sin 2\alpha}{\cosh 2k_0(s-\xi) - \cos 2\alpha}\right)\right.$$

$$+ \gamma(\xi)\cosh\ k_0(s-\xi)$$

$$\left.\times\left(\frac{1-\cos 2\alpha}{\cosh 2k_0(s-\xi) - \cos 2\alpha}\right)\right\};$$

$$[V](s) = \beta\,\frac{\sigma(s)}{2},$$

and $H(s - \xi)$ is the Heaviside step function (cf. (5.70)).

By following the arguments used in the previous section, the solution in the inner region can be written in the form:

$$\varphi^i \sim Us + F_1(s) \pm \frac{US'(s)}{2h}\,t + V_1(s)\,[t \pm C(s)],$$

(5.96)

where F_1 and V_1 are to be determined. Here S is the surface area and C is the blockage constant as before. This must be equal to (5.95) which leads to:

a. $\varphi_1(s) = F_1(s)$

b. $[\varphi_1](s) = V_1(s)C(s)$

c. $V + V(s) = V_1(s)$

d. $[V](s) = \dfrac{US'(s)}{h}.$

The last condition gives us the source strength in (5.94),

$$\sigma(s) = \frac{US'(s)}{\beta h}.$$

It remains now to determine $\gamma(\xi)$ from the other relations. Conditions (b) and (c) imply

$$\int_{-L/2}^{L/2} \gamma(\xi) \left\{ \left(\frac{1 - \cos 2\alpha}{\cosh 2k_0(s - \xi) - \cos 2\alpha} \right) \coth k_0(s - \xi) \right.$$

$$\left. + \frac{2w}{C(s)} H(s - \xi) \right\} d\xi$$

$$= - \frac{\pi U}{wh\beta^2} \sin 2\alpha \int_{-L/2}^{L/2} d\xi \frac{S(\xi) \sinh 2k_0(s - \xi)}{[\cosh 2k_0(s - \xi) - \cos 2\alpha]^2}$$

$$- 4wV := f(s).$$

This is an integral equation of the Cauchy type, therefore we write it in the following standard form (see (A.3)):

$$\frac{1}{\pi i} \int_{-L/2}^{L/2} \frac{d\xi\ \gamma(\xi)}{s - \xi} = F_\gamma(s), \qquad (5.97)$$

where

$$F_\gamma(s) := \frac{k_0}{\pi i} f(s) - \frac{2wk_0}{\pi i C(s)} \int_{-L/2}^{S} \gamma(\xi)\ d\xi$$

$$- \frac{1}{\pi i} \int_{-L/2}^{L/2} \gamma(\xi) \frac{A(s,\xi) k_0(s-\xi) \cosh k_0(s-\xi) - 1}{s - \xi} d\xi$$

with

$$A(s,\xi) := \frac{1 - \cos 2\alpha}{\cosh k_0(s - \xi) - \cos 2\alpha}.$$

From (5.97) we obtain a Fredholm integral equation of the second kind. The unique solution which obeys the Kutta condition at s = L/2 takes the form:

$$\gamma(s) = \frac{1}{\pi i} \sqrt{\frac{(s - L/2)}{(s + L/2)}} \int_{-L/2}^{L/2} \sqrt{\frac{(\zeta + L/2)}{(\zeta - L/2)}} \frac{F_\gamma(\zeta)}{x - \zeta} d\zeta.$$

By a similar argument as in Section 5.3, the sway force and yaw moment can be calculated:

$$F_2 = -\rho U h \int_{-L/2}^{L/2} \gamma(\xi)\, d\xi + \frac{\rho U}{2w\beta^2} \int_{-L/2}^{L/2} d\xi\, S'(\xi)$$

$$\times \int_{-L/2}^{L/2} d\zeta\, \gamma(\zeta) \frac{\sinh 2k_0(\xi - \zeta)}{\cosh 2k_0(\xi - \zeta) - \cos 2\alpha}$$

$$+ \frac{\rho U^2}{4w\beta^3 h} \sin 2\alpha \int_{-L/2}^{L/2} d\xi\, S'(\xi)$$

$$\times \int_{-L/2}^{L/2} d\zeta\, S'(\zeta) \frac{1}{\cosh 2k_0(\xi - \zeta) - \cos 2\alpha}$$

$$+ \frac{\rho h}{4w\beta} \sin 2\alpha \int_{-L/2}^{L/2} d\xi\, \gamma(\xi)$$

$$\times \int_{-L/2}^{L/2} d\zeta\, \gamma(\xi) \frac{1}{\cosh 2k_0(\xi - \zeta) - \cos 2\alpha},$$

$$F_6 = -\rho U h \int_{-L/2}^{L/2} \xi\gamma(\xi)\, d\xi - \frac{\rho U}{2w\beta^2} \int_{-L/2}^{L/2} d\xi\, S'(\xi)$$

$$\times \int_{-L/2}^{L/2} d\zeta\, \gamma(\zeta)$$

$$\times \frac{(\xi - \zeta) \sin^2 \alpha \coth k_0(\xi - \zeta) - \zeta \sinh 2k_0(\xi - \zeta)}{\cosh 2k_0(\xi - \zeta) - \cos 2\alpha}$$

$$+ \frac{\rho U^2 \sin 2\alpha}{8w\beta^3 h} \int_{-L/2}^{L/2} d\xi\, S'(\xi)$$

$$\times \int_{-L/2}^{L/2} d\zeta\, S'(\zeta) \frac{\xi + \zeta}{\cosh 2k_0(\xi - \zeta) - \cos 2\alpha}$$

$$+ \rho \frac{h \sin 2\alpha}{8w\beta} \int_{-L/2}^{L/2} d\xi\, \gamma(\xi)$$

$$\times \int_{-L/2}^{L/2} d\zeta\, \gamma(\zeta) \frac{\xi + \zeta}{\cosh 2k_0(\xi - \zeta) - \cos 2\alpha}.$$

To illustrate the idea, an example of calculation, carried out by Beck [1], based on these formulas is given in Figure 5.10 below. It turns out that the interaction terms between source and vortex distribution of the reflected points lead to large contributions in the sway force. These terms are asymptotically of lower order. However, due to summation, they become numerically of the same order.

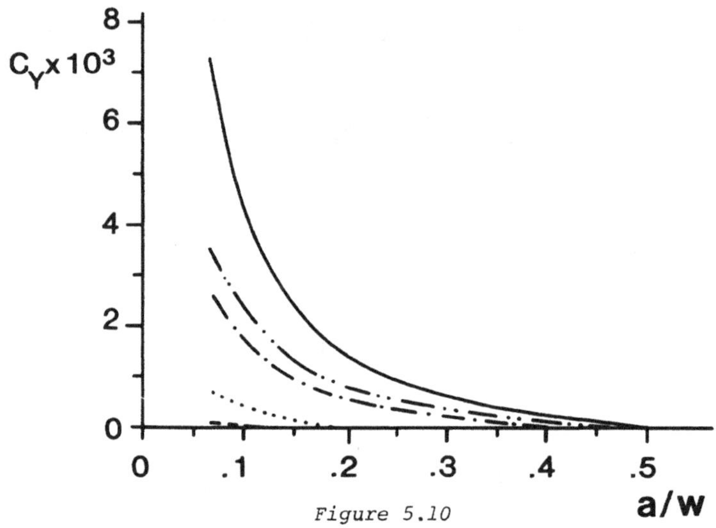

Figure 5.10

—— total; -..- first order; -.- second order, source term; .... second order, source-vortex term; --- second order, vortex term

5.5 Interaction of Ships

To conclude this chapter, we now consider the interaction of ships in shallow water. To simplify the presentation, we describe the situation as indicated in Figure 5.11. In particular, we assume that the lateral distance $S_p = O(1)$ as $\varepsilon \to 0$, where ε is the slenderness parameter. As in Figure 5.11, ship 1 travels at speed U_1 while the speed of the second ship is U_2. We consider a speed range for both ships such that $F_h^2 = U^2/gh = O(\varepsilon^2)$. Hence, outer potentials of both ships satisfy the Laplace equation with respect to a fixed coordinate system:

$$\varphi^0_{xx} + \varphi^0_{yy} = 0.$$

The outer solution may be considered as a source distribution along the center line of the ships together with a vortex distribution along the center line and the wakes of the ships. The source and vortex strength are time dependent. We have

$$\varphi^0(x,y,t) = \sum_{j=1}^{2} \frac{1}{2\pi} \int_{a_j^-(t)}^{a_j^+(t)} \sigma_j(\xi,t) \times \log \left[(x-\xi)^2 + \left(y - (-1)^j \frac{s_p}{2} \right)^2 \right]^{1/2} d\xi + \frac{1}{2\pi} \int_{-\infty}^{a_j^+(t)} \gamma_j(\xi,t) \arctan \left[\frac{y - (-1)^j \frac{s_p}{2}}{x-\xi} \right] d\xi. \tag{5.98}$$

We know that in the wake the vortices are transported along the stream line with the local velocity as in (5.5). This means

$$\frac{\partial \gamma_j}{\partial t}(x,t) = 0. \tag{5.99}$$

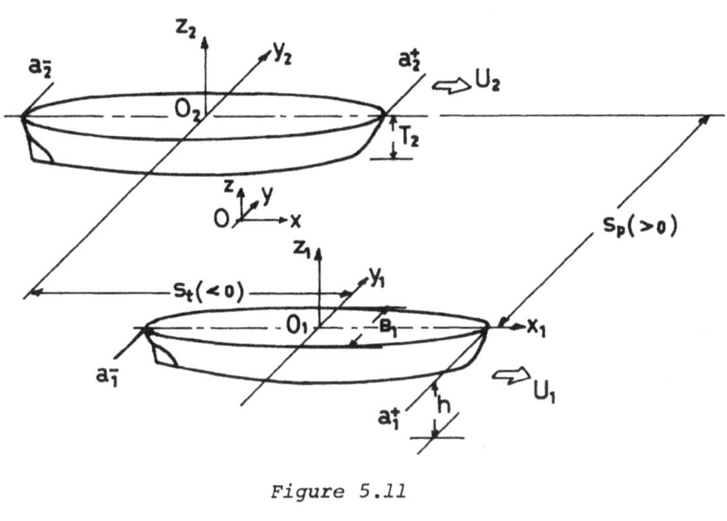

Figure 5.11

Furthermore, it follows from Kelvin's theorem that

$$\frac{d\Gamma_j}{dt} = -U_j \gamma_j\left(a_j^-(t)\right), \tag{5.100}$$

where Γ_j is the total circulation around the ship. To obtain uniqueness we require the Kutta condition:

$$\lim_{x_j \to a_j^- - 0} \gamma_j = \lim_{x_j \to a_j^- + 0} \gamma_j \tag{5.101}$$

holds for all t. Now we consider the field near ship 1 where (5.98) may be expanded in a Taylor series with respect to the ship coordinates (x_1, y_1):

$$\begin{aligned} \varphi^0(x_1, y_1, t) = {} & \varphi_1(x_1, \pm 0, t) \pm [\varphi_1](x, t) + V_{21}(x_1, t)\, y \\ & \pm \frac{\sigma_1(x_1, t)}{2} |y_1| + \frac{y_1}{2\pi} \int_{-\infty}^{a_1^+} \frac{\gamma_1(\zeta, t)}{x - \xi}\, d\xi + O(y_1^2), \\ & \text{for } a_1^- \leqslant x_1 \leqslant a_1^+ \end{aligned} \tag{5.102}$$

where

$$\begin{aligned} \varphi_1(x_1, \pm 0, t) = {} & \frac{1}{2\pi} \int_{a_1^-}^{a_1^+} \sigma_1(\xi, t) \log |x_1 - \xi|\, d\xi \\ & + \frac{1}{2\pi} \int_{a_2^-}^{a_2^+} \sigma_2(\xi, t) \log \left[(x_2 - \xi)^2 + s_p^2 \right]^{1/2} d\xi \\ & + \frac{1}{2\pi} \int_{-\infty}^{a_2^+} \gamma_2(\xi, t) \arctan\left(\frac{s_p}{x_2 - \xi} \right) d\xi, \end{aligned} \tag{5.103}$$

$$[\varphi_1](x_1, t) = \frac{1}{2} \int_{x_1}^{a_1^+} \gamma_1(\xi, t)\, d\xi,$$

and the induced normal velocity

$$V_{21}(x_1,t) := -\frac{1}{2\pi}\int_{a_2^-}^{a_2^+} \sigma_2(\xi,t)\,\frac{S_p}{(x_2-\xi)^2+S_p^2}\,d\xi$$

$$+\frac{1}{2\pi}\int_{-\infty}^{a_2^+} \gamma_2(\xi,t)\,\frac{x_2-\xi}{(x_2-\xi)^2+S_p^2}\,d\xi. \tag{5.104}$$

The solution in the inner region near ship 1 or 2 now again has an outer expansion

$$\varphi_j^i(y_j,z_j,x_j,t) = -\frac{(U_jS_j'(x_j))}{4h}|y_j|$$

$$+ V_j^*(x_j,t)\left[y_j \pm C_j(x_j)\right] + f_j(x_j,t).$$

Matching the inner and outer solution of ship 1 now leads to

$$\sigma_1(x_1,t) = -\frac{U_1S_1'(x_1)}{2h},$$

$$V_1^*(x_1,t) = V_{21}(x_1,t) + \frac{1}{2\pi}\int_{-\infty}^{a_1^+}\frac{\gamma_1(\xi,t)}{x_1-\xi}\,d\xi, \tag{5.105}$$

$$f_1(x_1,t) = \varphi_1(x,\pm 0,t),$$

$$V_1^*(x_1,t)C(x_1,t) = \frac{1}{2}\int_{x_1}^{a_1^+}\gamma_1(\xi,t)\,d\xi.$$

We obtain an integral equation for $\gamma_1(\xi,t)$ similar to (5.79):

$$\frac{1}{2\pi}\int_{-\infty}^{a_1^+}\left[\frac{\gamma_1(\xi,t)}{x_1-\xi} - \frac{\pi\gamma_1(\xi,t)}{C_1(x_1)}H(\xi-x_1)\right]d\xi = -V_{21}(x_1,t),$$

$$\text{for } a_1^- \leqslant x_1 \leqslant a_1^+. \tag{5.106}$$

In a similar way, we find near ship 2:

$$\frac{1}{2\pi}\int_{-\infty}^{a_2^+}\left[\frac{\gamma_2(\xi,t)}{x_2-\xi}-\frac{\pi\gamma_2(\xi,t)}{C_2(x_2)}\,H(\xi-x_2)\right]d\xi=-V_{12}(x_2,t),$$

$$\text{for } a_2^- \leqslant x_2 \leqslant a_2^+. \qquad (5.107)$$

Equations (5.106) and (5.107) are coupled singular integral equations for γ_1 and γ_2. To solve these equations, we may use a procedure with constant time step Δt. We start at a point $t = 0$ where no interaction takes place. This means that $\gamma_i = 0$ in the wake. We then start by taking V_{21} and V_{12} as the induced velocities due to the source distribution only. The integral equations may then be reduced to Fredholm equations with square integrable kernel (see Section 5.3) and can be solved numerically. The next time step gives us a constant vortex distribution along a wake with length $\Delta t U_j$ and the procedure can be repeated successively.

The sway force and yawing moment can be calculated by means of (5.43) and (5.44). An example of computation by Yeung [16] and experiments of Remery are shown in Figure 5.12. The computations are done by means of two different approximations of the solution of the integral equations.

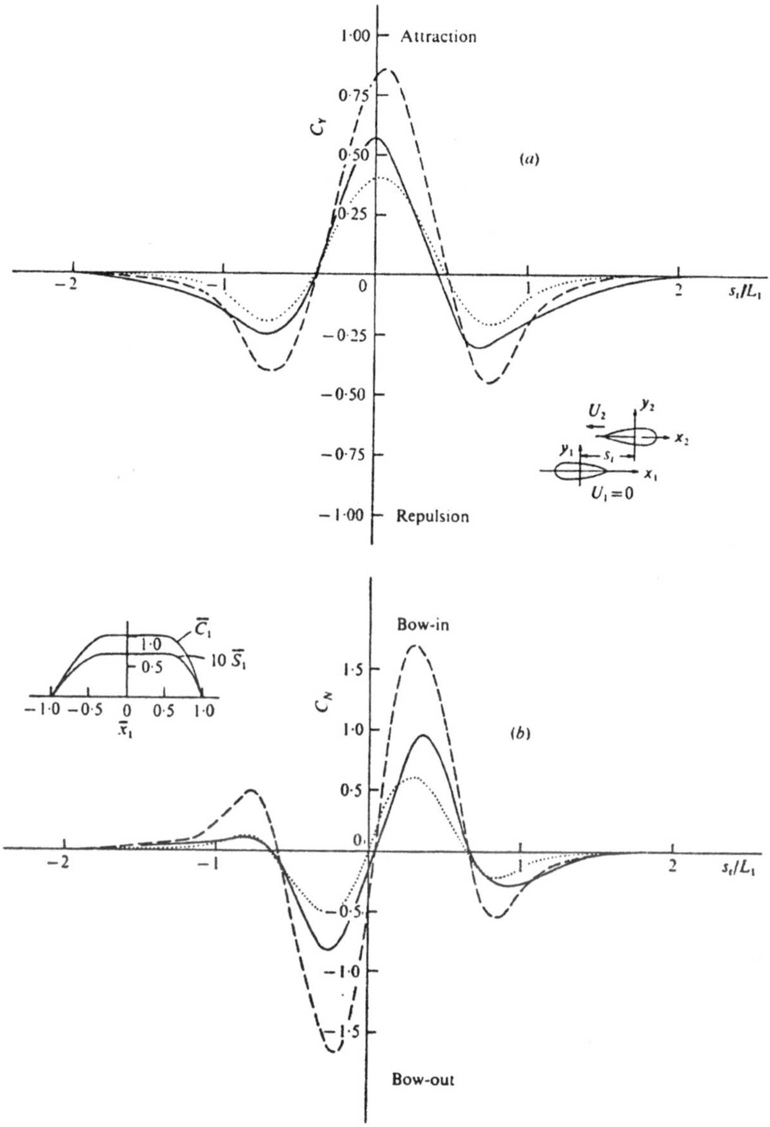

Figure 5.12

—— experiments

--- and ... computations

APPENDIX

In this appendix some mathematical tools, which are used in this book, are explained. For the asymptotic evaluation of single integrals with an oscillatory integrand the method of stationary phase is derived. Furthermore, the method of characteristics is explained which will be used to solve the Cauchy problem for first order partial differential equations.

Finally, for the asymmetric part of the flow problem, which results in the so-called Hilbert problems, we include here a short introduction to the theory of singular integral equations. Singular integral equations have been treated by Muskhelishvili [5] extensively.

A.1 The Method of Stationary Phase

This method dates back to Lord Kelvin, who developed it in particular for the case which is considered here. Suppose we have to evaluate an integral of the form:

$$I = \int_{\alpha}^{\beta} e^{i\omega f(x)}\; g(x)\; dx,$$

where f(x) and g(x) are differentiable real functions and ω is a large parameter. Integrating by parts, we find, if $f'(x) \neq 0$ in the interval $\alpha \leqslant x \leqslant \beta$,

$$\begin{aligned} I &= \frac{1}{i\omega} \int_{\alpha}^{\beta} \frac{g(x)}{f'(x)}\; d\left(e^{i\omega f(x)}\right) \\ &= \frac{1}{i\omega} \left[\frac{g(x)}{f'(x)}\; e^{i\omega f(x)}\right]_{\alpha}^{\beta} - \frac{1}{i\omega} \int_{\alpha}^{\beta} e^{i\omega f(x)}\; \frac{d}{dx} \left(\frac{g(x)}{f'(x)}\right) dx. \end{aligned}$$

Repeating this procedure as many times as differentiations permit, we have a series in inverse powers of ω, which gives an asymptotic expansion for I. Obviously, this method breaks down, if at a point γ in the interval $\alpha \leqslant x \leqslant \beta$, $f'(\gamma) = 0$. The point γ is usually referred to as the stationary point. For this case the method of stationary phase is developed.

Without loss of generality, let us consider an integral of the form:

$$J(0,x_0) = \int_0^{x_0} e^{i\omega f(x)} g(x)\, dx, \qquad x_0 > 0, \tag{A.1}$$

where $x = 0$ is the only stationary point of f. Here f and g have as many derivatives as we need, and again ω is a large parameter. At the origin f(x) now has an expansion

$$f(x) = f(0) + \frac{1}{2!} x^2 f''(0) + \frac{x^3}{3!} f'''(0) + \ldots$$

We assume that $f''(0) > 0$; if $f''(0) < 0$ the argument has to be slightly changed. We introduce a new variable z defined by

$$z = f(x) - f(0),$$

and consider x as a function of z. From the fact that $f'(0) = 0$, we see that x admits an expansion of the form:

$$x = z^{\frac{1}{2}} \sqrt{\frac{2}{f''(0)}} \left\{1 + c_1\sqrt{z} + c_2(\sqrt{z})^2 + \ldots\right\}.$$

In terms of the new variable z, the integral J takes the form:

$$e^{i\omega f(0)} \int_0^{f(x_0)-f(0)} e^{i\omega z} \left(g(x)\, \frac{dx}{dz}\right) dz. \tag{A.2}$$

Now if $g(x)\,\frac{dx}{dz}$ is expanded into powers of $\sqrt{z}$, we obtain

$$g(x)\, \frac{dx}{dz} = \frac{g(0)}{\sqrt{2f''(0)}}\, z^{-\frac{1}{2}} \left\{1 + \alpha_1\sqrt{z} + \alpha_2(\sqrt{z})^2 + \ldots\right\}, \tag{A.3}$$

which shows that we have to evaluate integrals of the form:

$$K_n(a) := \int_0^a e^{i\omega z} z^{(n-1)/2} dz, \qquad a = f(x_0) = f(0). \tag{A.4}$$

For real values of ω and a we cannot extend the upper limit to infinity. We calculate the integral $K_n(a)$ in the complex z-plane. Along a contour consisting of two vertical lines, joined by the interval $0 \leqslant z \leqslant a$, we see that

$$K_n(a) + \int_a^{a+i\infty} e^{i\omega z} z^{(n-1)/2} dz = \int_0^{i\infty} e^{i\omega z} z^{(n-1)/2} dz \tag{A.5}$$

by means of Cauchy's theorem. In the last integral in (A.5) we put $\omega z = i\lambda$ and transform it into:

$$\left(\frac{i}{\omega}\right)^{(n+1)/2} \int_0^{\infty} e^{-\lambda} \lambda^{(n-1)/2} d\lambda = \left(\frac{i}{\omega}\right)^{(n+1)/2} \Gamma\left(\frac{1}{2}(n+1)\right). \tag{A.6}$$

The second integral in (A.5) is transferred into

$$\left(\frac{i}{\omega}\right) a^{(n-1)/2} e^{i\omega a} \int_0^{\infty} e^{-\lambda} \left(1 + \frac{i\lambda}{a\omega}\right)^{(n-1)/2} d\lambda$$

by the substitution of $\omega z = \omega a + i\lambda$. Expanding $\left(1 + \frac{i\lambda}{a\omega}\right)^{(n-1)/2}$ into a power series in λ/ω,

$$1 + \frac{(n-1)}{2}\left(\frac{i}{a\omega}\right)\lambda + \frac{(n-1)(n-3)}{2^2 \cdot 2!}\left(\frac{i}{a\omega}\right)^2 \lambda^2 + \frac{(n-1)(n-3)(n-5)}{2^3 \cdot 3!}\left(\frac{i}{a\omega}\right)^3 \lambda^3 + \ldots,$$

we then obtain that

$$\left(\frac{i}{\omega}\right) a^{\frac{1}{2}(n-1)} e^{i\omega a} \int_0^\infty e^{-\lambda} \left(1 + \frac{i\lambda}{a\omega}\right)^{\frac{1}{2}(n-1)} d\lambda$$

$$= \left(\frac{i}{\omega}\right) a^{\frac{1}{2}(n-1)} e^{i\omega a}$$

$$\times \left\{1 + \frac{(n-1)}{2} \left(\frac{i}{a\omega}\right) \Gamma(2) + \frac{(n-1)(n-3)}{2^2\ 2!} \left(\frac{i}{\omega}\right)^2 \Gamma(3) + \ldots\right\}$$

$$= \left(\frac{i}{\omega}\right) a^{\frac{1}{2}(n-1)} e^{i\omega a}$$

$$\times \left\{1 + \frac{(n-1)}{2} \left(\frac{i}{a\omega}\right) + \frac{(n-1)(n-3)}{2^2} \left(\frac{i}{a\omega}\right)^2 + \ldots\right\}, \qquad (A.7)$$

by making use of the formula $\Gamma(n + 1) = \int_0^\infty e^{-\lambda} \lambda^n d\lambda = n!$ for integers n. Collecting results (A.5), (A.6), and (A.7), we have from (A.4)

$$K_n(a) = \left(\frac{i}{\omega}\right)^{\frac{1}{2}(n+1)} \Gamma\left(\frac{1}{2}(n + 1)\right) - \left(\frac{i}{\omega}\right) a^{\frac{1}{2}(n-1)} e^{i\omega a}$$

$$\times \left\{1 + \frac{(n-1)}{2} \left(\frac{i}{\omega a}\right) + \frac{(n-1)(n-3)}{2^2} \left(\frac{i}{\omega a}\right)^2 + \ldots\right\}$$

$$= \left(\frac{i}{\omega}\right)^{\frac{1}{2}(n+1)} \Gamma\left(\frac{1}{2}(n + 1)\right) - e^{i\omega a} \left(\frac{i}{\omega}\right)$$

$$\times \left\{\left(1 + \left(\frac{i}{\omega}\right) \frac{d}{da} + \left(\frac{i}{\omega}\right)^2 \frac{d^2}{da^2} + \ldots\right) a^{\frac{1}{2}(n-1)}\right\}. \qquad (A.8)$$

From (A.2), (A.7), and (A.8), it follows that the first order term of the asymptotic expansion (n = 0) of J in (A.1) takes the form:

$$J(0,x_0) = e^{i\omega f(0)} \frac{g(0)}{\sqrt{2f''(0)}} K_0(a)$$

$$= \frac{e^{i\omega f(0)}}{\omega^{\frac{1}{2}}} g(0) \sqrt{\frac{\pi i}{2f''(0)}} - \frac{ie^{i\omega f(x_0)}}{\omega}$$

$$\times \left\{\left(1 + \left(\frac{i}{\omega}\right) \frac{d}{da} + \left(\frac{i}{\omega}\right)^2 \frac{d^2}{da^2} + \ldots\right) \frac{g(0)}{\sqrt{2f''(0)}} a^{-\frac{1}{2}}\right\},$$

or

$$J(0,x_0) = \frac{e^{i\omega f(0)}}{\omega^{\frac{1}{2}}} g(0) \sqrt{\frac{\pi i}{2f''(0)}} - \frac{ie^{i\omega f(x_0)}}{\omega} \times \left(g(x)\, \frac{dx}{dz}\right)_{z=a} + \frac{e^{i\omega f(x_0)}}{\omega^2} \left(\frac{d}{dz}\left(g(x)\, \frac{dx}{dz}\right)\right)_{z=a} + \ldots, \tag{A.9}$$

from (A.3) with α_i being equal to zero. The first term being of order $\omega^{-\frac{1}{2}}$ is due to the stationary point and the following terms which have the forms resulted from the integration by parts are due to the end point of integration.

If we extend the interval from $z = a$ to $z = \beta$, we obtain:

$$e^{i\omega f(0)} \int_a^{\beta} e^{i\omega z} \left(g(x)\, \frac{dx}{dz}\right) dz$$

$$= e^{i\omega f(0)} \left[\frac{e^{i\omega z}}{i\omega} \left(g(x)\, \frac{dx}{dz}\right)\right]_a^{\beta} - \frac{e^{i\omega f(0)}}{i\omega} \int_a^{\beta} e^{i\omega z} \frac{d}{dz} \times \left(g(x)\, \frac{dx}{dz}\right) dz, \tag{A.10}$$

the contribution from the interval where $f'(x)$ is not equal to zero. Adding the two integrals from (A.9) and (A.10), we have

$$J(0,\beta) = \frac{e^{i\omega f(0)}}{\omega^{\frac{1}{2}}} g(0) \sqrt{\frac{\pi i}{2f''(0)}} - \frac{ie^{i\omega f(\beta)}}{\omega} \times \left(g(x)\, \frac{dx}{dz}\right)_{z=\beta} + O(1/\omega), \tag{A.11}$$

where the term of $O(1/\omega)$ is due to K_1. Hence, for large ω, the stationary point gives the largest contribution (order $\omega^{-\frac{1}{2}}$) while the contribution from the end points is of lower order ω^{-1}. Similarly, it is easy to see that

$$J(\alpha,0) = \frac{e^{i\omega f(0)}}{\omega^{\frac{1}{2}}} g(0) \sqrt{\frac{\pi i}{2f''(0)}} + \frac{ie^{i\omega f(\alpha)}}{\omega} \left(g(x)\, \frac{dx}{dz}\right)_{z=\alpha} + O(1/\omega). \tag{A.12}$$

From (A.11) and (A.12), we see that if $x = \gamma$ is the only stationary point of f in the interval $\alpha \leqslant x \leqslant \beta$, then

$$J(\alpha,\beta) = \frac{e^{i\omega f(\gamma)}}{\omega^{\frac{1}{2}}} g(\gamma) \sqrt{\frac{2\pi i}{f''(\gamma)}} = O(1/\omega). \qquad (A.13)$$

A.2 Characteristics of First Order Partial Differential Equations

The equation

$$F(x,y,u,u_x,u_y) = 0, \qquad (A.14)$$

defines a relation between the function $u = u(x,y)$ and its derivative u_x and u_y. In order to define the function u as a solution of this equation, a set of subsidiary conditions must be prescribed. A fundamental problem for (A.14) is the *Cauchy problem* which consists in finding an integral surface $u(x,y)$ of the equation (A.14) passing through a prescribed curve C in the 3-dimensional space given by

$$x = x(\lambda), \quad y = y(\lambda), \quad u = u(\lambda), \qquad (A.15)$$

where x,y,u are smooth functions of the parameter λ.

Consider first the case that equation (A.14) is a quasi-linear equation, i.e.,

$$F = Au_x + Bu_y + C = 0, \qquad (A.16)$$

where A, B and C are functions of x, y and u. If the Cauchy problem has a unique solution, the derivatives u_x and u_y along the curve C become functions of the parameter λ and must be determined uniquely by the two conditions:

$$Au_x + Bu_y + C = 0, \text{ and}$$

$$u_x \frac{dx}{d\lambda} + u_y \frac{dy}{d\lambda} - \frac{du}{d\lambda} = 0.$$

This is possible only if $\frac{\frac{dx}{d\lambda}}{A} \neq \frac{\frac{dy}{d\lambda}}{B}$. If, however,

$$\frac{\frac{dx}{d\lambda}}{A} = \frac{\frac{dy}{d\lambda}}{B} = \frac{-\frac{du}{d\lambda}}{C}, \tag{A.17}$$

the system admits a one-parameter family of solutions, i.e., the solution to the Cauchy problem (if it exists) is not unique. The curves that satisfy (A.17) are called the *characteristic* curves of the differential equation (A.16). Through a characteristic curve, several solutions of (A.16) may pass, they can, in many cases, be considered as lines of propagation of small disturbances (discontinuities in the derivatives of u). It is easily shown that the Cauchy problem for a nowhere characteristic curve C can be solved by constructing the solutions to the characteristic equations at each point of the curve and by considering the surface, $u = u(x,y)$, which is obtained in this way.

For a non-linear partial differential equation (A.14), the definition of a characteristic curve is more complicated. At a fixed point $P = (x,y,u)$ the equation (A.14) determines a relation between u_x and u_y, i.e., it determines one-parameter family of the possible tangent planes to the integral surfaces, the solutions of (A.14). The envelope of these planes forms a cone with vertex at P, which is known as the *Monge cone*.

Consider again a curve C such as defined by (A.15). Then at a point on C the equation (A.14) and the compatibility condition,

$$u_x \frac{dx}{d\lambda} + u_y \frac{dy}{d\lambda} = \frac{du}{d\lambda}, \tag{A.18}$$

define a number of possible tangent planes. If we assume that F is not factorable, we cannot have dependence; hence the first order derivatives u_x and u_y are always determined. The characteristic curves are obtained by considering a breakdown of the determination of the *second-order* derivatives. Differentiating (A.18) with respect to λ yields:

$$u_{xx}\left(\frac{dx}{d\lambda}\right)^2 + 2u_{xy}\left(\frac{dx}{d\lambda}\right)\left(\frac{dy}{d\lambda}\right) + u_{yy}\left(\frac{dy}{d\lambda}\right)^2$$

$$= \frac{d^2u}{d\lambda^2} - u_x \frac{d^2x}{d\lambda^2} - u_y \frac{d^2y}{d\lambda^2}, \tag{A.19}$$

and differentiating (A.14) with respect to x and y yields:

$$\begin{aligned} u_{xx}F_p + u_{xy}F_q + F_u p + F_x &= 0; \\ u_{xy}F_p + u_{yy}F_q + F_u q + F_y &= 0, \end{aligned} \tag{A.20}$$

with $p = u_x$ and $q = u_y$. As shown before, p and q can be uniquely determined as function of λ. Thus, it follows from differentiation that

$$\begin{aligned} \frac{dp}{d\lambda} &= u_{xx}\frac{dx}{d\lambda} + u_{xy}\frac{dy}{d\lambda}; \\ \frac{dq}{d\lambda} &= u_{xy}\frac{dx}{d\lambda} + u_{yy}\frac{dy}{d\lambda}. \end{aligned} \tag{A.21}$$

Equations (A.20) and (A.21) give a set of four equations for the three unknowns, u_{xx}, u_{xy} and u_{yy}. The augmented matrix A,

$$\begin{pmatrix} F_p & F_q & 0 & pF_u+F_x \\ 0 & F_p & F_q & qF_u+F_y \\ \frac{dx}{d\lambda} & \frac{dy}{d\lambda} & 0 & -\frac{dp}{d\lambda} \\ 0 & \frac{dx}{d\lambda} & \frac{dy}{d\lambda} & -\frac{dq}{d\lambda} \end{pmatrix},$$

will have rank generally equal to three. If it has a rank equal to two, then the curve C defined by (A.15) is a *characteristic*. In this case all the 3 × 3 submatrices of A are singular. This leads to the Charpit-Lagrange equations:

$$\frac{dx}{F_p} = \frac{dy}{F_q} = \frac{du}{pF_p + qF_q} = -\frac{dp}{F_x + pF_u} = -\frac{dq}{F_y + qF_u}, \tag{A.22}$$

for the quantities x, y, u, p and q along a characteristic. Equations (A.22) are also known as the *characteristic equations* for the differential equation (A.14). Here again for a non-characteristic curve, the Cauchy problem is solved by constructing a surface, which is described by the characteristics issued from the points of this curve. (Note that the initial values of p and q at the point are determined!)

A.3 Singular Integral Equations

Let $\mathcal{L}$ be a segment of a closed smooth contour C in the complex z-plane. Let S^+ and S^- denote, respectively, the regions inside and outside the contour C (see Figure A.1). Consider the integral along $\mathcal{L}$

$$H(z) = \frac{1}{2\pi i}\int_{\mathcal{L}} \frac{h(s)}{s - z}\, ds, \tag{A.23}$$

where h(s) is a given Hölder continuous function (complex) along $\mathcal{L}$ such that

$$|h(s_2) - h(s_1)| \leqslant A|s_2 - s_1|^{\mu},$$

$$\text{for } s_1,\ s_2 \text{ on } \mathcal{L},$$

with positive constants A and μ.

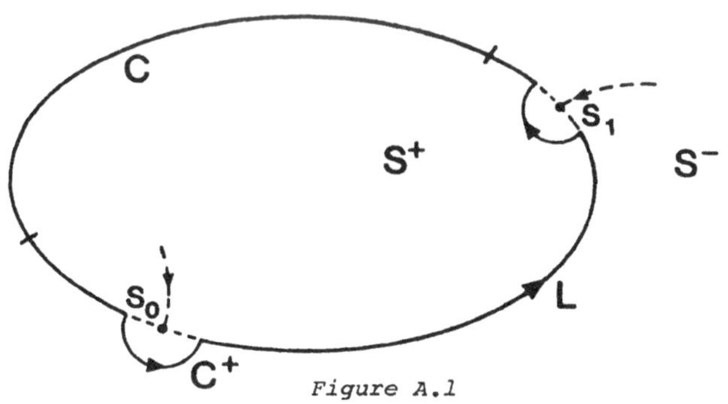

Figure A.1

Let s_0 be a fixed point on $\mathcal{L}$. Then we have

$$\lim_{\substack{z\to s_0 \\ z\in S^+}} H(z) = \frac{1}{2\pi i}\int_{\mathcal{L}} \frac{h(s)}{s - s_0}\, ds + \frac{1}{2\pi i}\int_{C^+} \frac{h(s)}{s - s_0}\, ds,$$

where $\fint$ denotes the principle value, and C^+ is a semicircle in S^- with radius ρ. The integral along C^+ can be worked out by means of a Taylor series expansion of $h(s)$ at the point s_0. We have

$$\frac{1}{2\pi i}\int_{C^+}\frac{h(s)}{s - s_0}\,ds = \frac{1}{2\pi i}\int_{C^+}\left\{\frac{h(s_0)}{s - s_0} + h'(s_0) + \ldots\right\} ds.$$

By introducing $s - s_0 = \rho e^{i\vartheta}$, the integral can be rewritten as:

$$\frac{1}{2\pi}\int_{\pi}^{2\pi}\left\{h(s_0) + \rho e^{i\vartheta}\, h'(s_0) + \ldots\right\} d\vartheta.$$

Hence if z approaches $\mathcal{L}$ from S^+ and S^- we obtain, respectively

$$H^+(s_0) = \frac{1}{2}\,h(s_0) + \frac{1}{2\pi i}\fint_{\mathcal{L}}\frac{h(s)}{s - s_0}\,ds, \tag{A.24}$$

and

$$H^-(s_0) = -\,\frac{1}{2}\,h(s_0) + \frac{1}{2\pi i}\fint_{\mathcal{L}}\frac{h(s)}{s - s_0}\,ds. \tag{A.25}$$

These are the well-known Plemelj formulas from which we see that

$$\begin{aligned} H^+(s_0) - H^-(s_0) &= h(s_0) \\ H^+(s_0) + H^-(s_0) &= \frac{1}{\pi i}\fint_{\mathcal{L}}\frac{h(s)}{s - s_0}\,ds. \end{aligned} \tag{A.26}$$

Consider now the relation

$$H^+(s) = G(s)\,H^-(s) + g(s), \qquad \text{on } \mathcal{L}, \tag{A.27}$$

where G and g are given functions on $\mathcal{L}$. We are interested here in the determination of $H(z)$ as an analytic continuation of the function H^+ and H^- on $\mathcal{L}$. This is so-called the *Hilbert problem*. This can be done easily by introducing a function $\chi(z)$ which obeys the homogeneous relation

$$G(s) = \frac{\chi^+(s)}{\chi^-(s)}, \qquad \text{on } \mathcal{L},$$

from which it follows that

$$\frac{H^+(s)}{\chi^+(s)} - \frac{H^-(s)}{\chi^-(s)} = \frac{g(s)}{\chi^+(s)}.$$

Thus, if

$$H(z) = \frac{\chi(z)}{2\pi i}\left\{\int_{\mathcal{L}} \frac{g(s)}{\chi^+(s)(s - z)}\, ds + P(z)\right\}, \qquad (A.28)$$

then H(z) is the corresponding analytic continuation for any arbitrary analytic function P(z).

REFERENCES

[1] R.F. Beck, Forces and Moments on a Ship Moving in a Shallow Channel, J. of Ship Res., 21, 1977.

[2] S. Bochner, Vorlesungen über Fouriersche Integrale, New York, Chelsea Publishing Company, 1948.

[3] H.W. Hoogstraten, On Non-linear Dispersive Water Waves, Thesis, Delft University of Technology, 1968.

[4] Lord W.T. Kelvin, On the Waves Produced by a Single Impulse in Water of any Depth, or in a Dispersive Medium, Proceedings of the Royal Society of London, Ser. A, Vol. 42, 1887.

[5] N. Muskhelishvili, Singular Integral Equations, Wolters & Noordhoff, 1953.

[6] J.N. Newman, Lateral Motion of a Slender Body Between Two Parallel Walls, J. Fl. Mech., 1969.

[7] T.F. Ogilvie, Recent Progress Toward the Understanding and Prediction of Ship Motion, Fifth Symposium, Naval Hydrodynamics, ACR-112, ONR, 1964.

[8] J.J. Stoker, Water Waves, New York, Interscience, 1957.

[9] R. Timman and J.N. Newman, The Coupled Damping Coefficients of a Symmetric Ship, Journal of Ship Research, Vol. 5, No. 4, 1962.

[10] E.O. Tuck, Shallow Water Theory Past Slender Bodies, J. Fl. Mech., 26, 1966.

[11] M. van Dyke, Perturbation Methods in Fluid Mechanics, Acad., 1964.

[12] J.V. Wehausen and E.V. Laitone, Surface Waves, Handbuch der Physik, Vol. IX, Berlin, Springer-Verlag, 1960.

[13] N. Wiener, The Fourier Integral and Certain of Its Applications, New York, Dover, 1933.

[14] L.C. Woods, The Theory of Subsonic Plane Flow, Cambridge University Press, 1961.

[15] R.W. Yeung, Applications of Slender Body Theory of Ships Moving in Restricted Shallow Water, Proc. of Symp. on Aspects of Navigability, Delft, 1978.

[16] R.W. Yeung, On the Interaction of Slender Ships in Shallow Water, J. of Fl. Mech., 85, 1978.

INDEX

MIX
Papier | Fördert gute Waldnutzung
FSC® C083411

Zeitfracht Medien GmbH
Ferdinand-Jühlke-Straße 7
99095 Erfurt, Deutschland
produktsicherheit@kolibri360.de